Physiology of Biodegradative Microorganisms

Physiology of Biodegradative Microorganisms

edited by

Colin Ratledge

Department of Applied Biology, University of Hull, UK

Reprinted from *Biodegradation* 1: 2/3, 1990

Kluwer Academic Publishers

Dordrecht / Boston / London

Library of Congress Cataloging-in-Publication Data

Physiology of biodegradative microorganisms / edited by C. Ratledge.
p. cm.
ISBN 0-7923-1132-9 (alk. paper)
1. Microbial metabolism. 2. Biodegradation. I. Ratledge, Colin.
QR88.P48 1991
576'.11--dc20 91-6305

ISBN 0-7923-1132-9

Published by Kluwer Academic Publishers,
P.O. Box 17, 3300 AA Dordrecht, The Netherlands.

Kluwer Academic Publishers incorporates
the publishing programmes of
D. Reidel, Martinus Nijhoff, Dr W. Junk and MTP Press.

Sold and distributed in the U.S.A. and Canada
by Kluwer Academic Publishers,
101 Philip Drive, Norwell, MA 02061, U.S.A.

In all other countries, sold and distributed
by Kluwer Academic Publishers Group,
P.O. Box 322, 3300 AH Dordrecht, The Netherlands.

Printed on acid-free paper

Printed in The Netherlands

Contents

C. Ratledge (ed.) Physiology of Biodegradative Microorganisms. 1: VII–VIII.

Editorial

It has been said that without microorganisms animal life on the planet would cease to exist within about five years. Whether or not this is an exaggeration in time-scale, it is clearly true in principle that we depend absolutely on the activities of microorganisms for the replenishment of our environment. Over the years we have built up a shrewd idea of how the major components of the biological world are degraded, though there are still some surprising omissions such as a lack of detail for the breakdown of DNA, RNA and their component nucleotide bases. There is still much to be learnt about the degradation of complex ligno-cellulosic materials, particularly those found in some of the hardwoods, which can be very resistant to decay.

The concept of microroganisms as an essential part of the recycling process is, of course, not new. However, the ability of microorganisms to deal with man-made chemicals – the xenobiotic materials – has, of course, been of more recent occurrence. It was some 30 to 35 years ago that microbiologists first became seriously interested in solving some of the challenges being offered by the molecules produced by synthetic chemists. The ability of microorganisms to break down a relatively large number of synthetic aromatic compounds has been known, at least in principle, for much longer, but the pathways for the degradation of benzenoid or naphthalenic aromatic compounds were obscure. Now, thanks to a sustained effort by many research groups over the past three of four decades, we have an accurate understanding of how the majority of such compounds are degraded.

However, knowing the details of a pathway, or even an array of interlocking pathways, is not sufficient in itself. Knowledge should always be leading us somewhere. The knowledge gained in my example of aromatic compound degradation is now standing us in extremely good stead as it is providing the basis from which we can begin to tackle some of the major outstanding problems of biodegradation, namely the elimination of the persistent recalcitrant molecules which the chemical industry has been manufacturing for a wide variety of activities.

This is not the place to enter into the controversy of whether such compounds do more good than harm: the fact remains that materials are used and some, especially halogenated derivatives, are extremely resistant to microbial (and therefore to all biological) breakdown. Again, I do not want to develop the argument that such compounds, by their very non-degradability, are remarkably unreactive, so that they are not likely to represent a toxicological threat to man, his animals or the environment. Whatever the advantages or disadvantages of using such compounds may be, they do exist and when used, do persist. To deal with them effectively calls for the combined skills of the microbial biochemist and geneticist.

To deal with man-made chemicals it may be necessary to use man-made microorganisms, or at least to produce genetically modified enzymes which can now attack some of the previously recalcitrant molecules. Such a programme, which is exemplified by current activities to deal with the degradation of the poly-chlorobenzenes and biphenyls, can succeed only because we have already acquired the essential knowledge as to how the parent molecules of benzene and biphenyl are themselves degraded. This principle extends to all other types of molecules.

The ability to define the route to biodegradability of a given molecule is now here, and there are many examples of how such knowledge has successfully been applied. Perhaps the best example of this is in the redesigning of the formulation of lubricating oils going into two-stroke outboard motor-boat engines. Knowing what would constitute a biodegradable molecule, it was possible to formulate an entirely bio-degradable oil which could be used by boats operating in inland lakes. Build-up of potentially harmful compounds in the lake was prevented and the ecosystem was not then challenged in such a way that it may not have been able to withstand.

Our knowledge of how biodegradation proceeds is vital because biodegradation itself is vital. In this series of contributions, I hope we present sufficient breadth for the reader to see that there are few areas that are escaping the attentions of the microbial physiologist. The coverage which is assembled here is, however, not a complete one and was never intended to be so. It gives a cross-section – and I hope a fairly typical cross-section – of what are some of the major points of microbial research today on this area.

There are many materials which suffer from unwanted biodegradation. Sometimes our knowledge can be used to prevent this happening but it must be said that there are many areas where the basic information about degradative pathways is unknown. For such compounds, degradation is an empirical study. I would though, hope that through these present articles the reader will be able to gather that on many key areas we have moved from empiricism into a rational understanding of the biochemistry of degradation.

I would like to thank all the contributors to this Special Issue for responding, not only as to the initial invitation to contribute but also to meet a very tight deadline by which their contribution had to be completed. We lost only one contribution which was to have covered degradation of fatty acids, as the author, unfortunately for us, moved his laboratory from one continent to another just at the critical time.

To all the contributing authors, I would offer my sincere thanks for their endeavours. I hope that you the reader, will find the chapters illuminating, instructive, but, not least, a justification for believing that microbial biochemistry is indeed the true cornerstone for the understanding of biodegradation.

University of Hull, UK Colin Ratledge

Biodegradation **1**: 79–92, 1990.

Physiology of aliphatic hydrocarbon-degrading microorganisms

Robert J. Watkinson & Philip Morgan*
Shell Research Ltd., Sittingbourne Research Centre, Sittingbourne, Kent, ME9 8AG, UK
(requests for offprints)*

Key words: aliphatic hydrocarbons, alkanes, alkenes, biodegradation, metabolism

Abstract

This paper reviews aspects of the physiology and biochemistry of the microbial biodegradation of alkanes larger than methane, alkenes and alkynes with particular emphasis upon recent developments. Subject areas discussed include: substrate uptake; metabolic pathways for alkenes and straight and branched-chain alkanes; the genetics and regulation of pathways; co-oxidation of aliphatic hydrocarbons; the potential for anaerobic aliphatic hydrocarbon degradation; the potential deployment of aliphatic hydrocarbon-degrading microorganisms in biotechnology.

Introduction

Aliphatic hydrocarbons represent a wide range of potential substrates for microorganisms. They may be saturated (alkanes) or unsaturated (alkenes and alkynes). They range from gases, such as methane and ethane, through liquids to long-chain molecules of 40 or more carbon atoms that are solid at physiological temperatures. They may be straight-chain compounds, simple branched compounds or highly branched. However, they are all insoluble, hydrophobic molecules composed entirely of carbon-carbon and carbon-hydrogen linkages.

A wide variety of bacteria, filamentous fungi and yeasts can metabolize aliphatic hydrocarbon substrates. Although the identities assigned to organisms in older papers may be of questionable validity to the modern taxonomist, it is important to be aware of the diversity of microorganisms capable of degrading alkanes and alkenes. Table 1 is a partial list of genera of microorganisms that have been shown to metabolize aliphatic hydrocarbons. The physiology of microbial aliphatic hydrocarbon degradation has been extensively studied. From a fundamental viewpoint there has been particular interest in substrate uptake mechanisms and the metabolic processes responsible for initiating catabolism. There is also extensive interest in the application of aliphatic hydrocarbon-degraders in biotechnology. In a review such as this it is only possible to consider selected examples of physiology in order to illustrate the fundamental properties of microbial utilisation of these substrates. It is the aim of this paper to review recent developments in our understanding of the physiology and biochemistry of aliphatic hydrocarbon degradation and to relate fundamental knowledge to the potential applications of such organisms. A specific exception to the coverage will be the degradation of methane since the metabolism of this compound appears to represent a relatively specialised physiology confined to a distinct group of microorganisms (de Vries et al. 1990).

Microbial uptake of aliphatic hydrocarbons

Physicochemical properties of aliphatic hydrocarbons

Aliphatic hydrocarbons pose a variety of challenges to degradative microorganisms due to their fundamental physicochemical properties. Table 2 lists some basic properties of a few selected examples of n-alkanes, branched alkanes and alkenes. The physical state of the compounds at physiological temperatures may be gaseous, liquid or solid. It is generally true to state that the gaseous and liquid compounds are the most readily degraded but liquids of lower molecular weight may prove to be inhibitory to microorganisms by virtue of their solvent effect (Atlas 1981; Pfaender & Buckley 1984). However, the most significant property of aliphatic hydrocarbons with respect to their utilisation as metabolic substrates is their extremely limited solubility in water. As can be seen from Table 2, the solubility of aliphatic compounds rapidly decreases with increasing molecular weight. From a microbiological viewpoint, the solubility of aliphatic hydrocarbons can be considered as insignificant except for the compounds of very low molecular weight. As a consequence, microorganisms have had to develop a variety of specific adaptations in order to be able to utilise the majority of potential hydrocarbon substrates. These will be briefly discussed below and are, of course, equally relevant for the utilisation of other poorly soluble substrates.

Microbial adaptations for hydrocarbon uptake

The challenges to substrate uptake presented by the insolubility and hydrophobicity of aliphatic hydrocarbons may be met by microorganisms in a variety of ways. There are three possible routes for hydrocarbon uptake: soluble materials only; *via* microdroplets (i.e. droplets much smaller than the microbial cell); *via* macrodroplets. The uptake of aliphatic hydrocarbons following their dissolution in water is only feasible for the very low molecular weight compounds since the heavier compounds exhibit both negligible solubility and slow dissolution. This problem was elegantly modelled by Miller & Bartha (1989) who demonstrated uptake-limitation of the degradation of n-hexatricontane (nC_{36}) in aqueous cultures by comparing degradation rates before and after microencapsulation of the hydrocarbon substrate.

The uptake of hydrocarbon in droplet form is very common and frequently involves the production of biological surfactant molecules as emulsifying agents to produce microdroplets of hydrocarbon. In such cases the hydrocarbon droplets may be encapsulated within a surfactant micelle. Biological surfactants are described in detail elsewhere in this issue (see paper by Hommel). Droplets of any size or free-phase hydrocarbon material must then be taken up by the microorganisms and there exist a variety of adaptations to facilitate this. This has been extensively reviewed by Finnerty & Singer (1985). Many hydrocarbon-degrading microorganisms have highly hydrophobic cell surfaces and may frequently associate with hydrocarbon droplets or pass into the organic phase during growth. It has been widely demonstrated that extensive changes in membrane lipid composition occur during growth on alkanes (Singer & Finnerty

Table 1. Some genera of microorganisms that have been shown to metabolize aliphatic hydrocarbons other than methane.

Bacteria	Yeasts	Filamentous fungi
Acetobacter	Candida	Aspergillus
Acinetobacter	Cryptococcus	Cladosporium
Actinomyces	Debaryomyces	Corollaspora
Alcaligenes	Hansenula	Dendryphiella
Bacillus	Pichia	Gliocladium
Beneckea	Rhodotorula	Lulworthia
Corynebacterium	Sporobolomyces	Penicillium
Flavobacterium	Torulopsis	Varicospora
Mycobacterium	Trichosporon	
Nocardia		
Pseudomonas		
Rhodococcus		
Xanthomonas		

Selected information from Britton (1984); van Ginkel & de Bont (1986); Hommel & Kleber (1984); Kirk & Gordon (1988); Lindley et al. (1986); Nakajima et al. (1985); Wood & Murrell (1989).

1984; Ratledge 1978). In some cases this may represent an adaptation for cell association with the hydrocarbon phase. Ng & Hu (1989) showed that the production of biosurfactants by *Acinetobacter calcoaceticus* did not affect the association of the cells with the hydrocarbon phase. Microscopic studies of yeasts give evidence for pores in the cell wall that permit the penetration of hydrocarbons to the surface of the cell membrane (Scott & Finnerty 1976a). Transport across the membrane has generally been thought to be a passive process (Ratledge 1978) but there is evidence for an energy requirement for uptake in some yeasts (Scott & Finnerty 1976b; Bassel & Mortimer 1985). In either case intracellular hydrocarbon droplets can then be observed microscopically (Scott & Finnerty 1976a,b).

A particularly interesting bacterial system has been described by Kappeli & Finnerty (1979) for *Acinetobacter* HO1-N growing on n-hexadecane. This organism solubilises the hydrocarbon by encapsulating it in membrane microvesicles which are then taken into the cell by an active process (Singer & Finnerty 1984a). Intracellular structures have been observed in other organisms growing on hydrocarbons. For example, Watkinson (1980) described intracellular vesicles and tubules in a *Nocardia* which were suggested to play a role in hydrocarbon uptake. It should be noted, however, that intracellular inclusions produced during growth on hydrocarbons may be a consequence of hydrocarbon utilisation rather than an adaptation for uptake. Such structures are exemplified by microbodies in alkane-grown yeasts which are structures rich in oxidative enzymes (Fukui & Tanaka 1979). Indeed, Hommel & Ratledge (1990) showed that the fatty alcohol oxidases involved in n-alkane metabolism by *Candida bombicola* were solely present in the microsomal fraction. Whilst the details of hydrocarbon uptake systems in many yeasts and bacteria have yet to be elucidated, even less information is available concerning hydrocarbon uptake by filamentous fungi. Kirk & Gordon (1988) have demonstrated that some marine, alkane-degrading, filamentous fungi produce emulsifying agents which cause the production of hydrocarbon droplets. These are then surrounded and penetrated by hyphae.

Metabolic pathways for aliphatic hydrocarbons

The metabolic pathways responsible for the degradation of a wide variety of hydrocarbons have been reported in the literature. Generally, degradation can only be initiated under aerobic conditions since oxygenase reactions appear to be necessary for the initial metabolic activation of alkane molecules. Some reports of anaerobic degradation have been made and these will be critically discussed subsequently. Earlier literature tended to emphasise

Table 2. Physical properties of selected aliphatic hydrocarbons.

Compound	C atoms	Mol. wt.	m.p. (°C)	b.p. (°C)	Solubility (mg l^{-1})
Ethane	2	30.1	− 172.0	− 88.6	63.7
n-hexane	6	86.2	− 94.3	68.7	12.3
n-decane	10	128.3	− 31.0	174.0	0.05
n-hexadecane	16	226.4	19.0	287.0	5.2×10^{-5}
n-eicosane	20	282.6	36.7	343.0	3.1×10^{-7}
n-hexacosane	26	366.7	56.4	412.2	1.3×10^{-10}
2-methylpentane	6	86.2	− 154.0	60.3	13.8
2,2,4-trimethylpentane	8	114.2	− 107.2	127.0	2.4
4-methyloctane	9	128.3	–	142.0	0.12
1-hexene	6	84.2	− 139.8	63.5	50.0
trans-2-heptene	7	98.2	− 109.5	98.0	15.0
1-octene	8	112.2	− 121.3	121.0	2.7

Data selected from Eastcott et al. (1988). m.p. is the melting point and b.p. the boiling point at normal temperature and pressure.

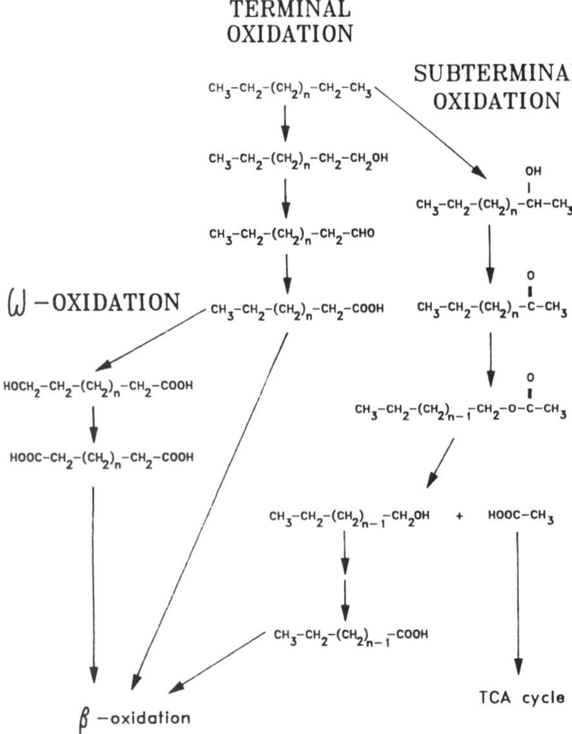

Fig. 1. Basic metabolic pathways for the degradation of n-alkanes. Illustrated are the three main metabolic routes documented for microorganisms: terminal oxidation; terminal oxidation followed by ω-oxidation; subterminal oxidation.

degradation of n-alkanes and, to a lesser extent, simple branched alkanes. This material has been extensively reviewed, for example by Ratledge (1978), Britton (1984) and Singer & Finnerty (1984a). Consequently, it is only necessary for the purposes of this review to reiterate briefly the basic principles of these metabolic pathways. Research described in recent years has resulted in a greater understanding in the metabolism of more complex branched alkanes, of alkenes and of co-metabolism of aliphatic compounds.

Metabolism of n-alkanes

Of the aliphatic hydrocarbons, it is the n-alkanes that are claimed to be the most rapidly degraded components in both laboratory culture and the natural environment (Wakeham et al. 1986; Kennicutt 1986; Oudot et al. 1989). The fundamental details of n-alkane degradation have been well documented (Britton 1984; Singer & Finnerty 1984a). Most microorganisms convert n-alkanes to the corresponding alkan-1-ol by means of a hydroxylase (monooxygenase) system:

$$R\text{-}CH_3 + O_2 + NAD(P)H + H^+ \rightarrow R\text{-}CH_2OH + NAD(P)^+ + H_2O$$

Hydroxylation reactions of this type may be linked to a number of types of electron carrier system. Systems linked to rubredoxin (e.g. *Pseudomonas putida*) and cytochrome P-450 (e.g. *Candida* spp.) have been the most thoroughly investigated. Dioxygenase systems have also been reported but these are less common. In these systems the n-alkanes are transformed into the corresponding hydroperoxides and subsequently reduced to the corresponding alkan-1-o1:

$$R\text{-}CH_3 + O_2 \rightarrow R\text{-}CH_2OOH + NAD(P)H + H^+ \rightarrow R\text{-}CH_2OH + NAD(P)^+ + H_2O$$

Subterminal oxidations of n-alkanes to secondary alcohols may also occur but this is rarer. For example, Rehm & Reiff (1982) investigated the initial mode of attack on C_8 to C_{18} n-alkanes by a variety of bacteria and fungi. Most organisms brought about terminal oxidation only but with certain *Aspergillus, Fusarium* and *Bacillus* spp. subterminal oxidation was detected. This resulted primarily in the production of 4-, 5-, or 6-substituted products with lesser amounts of 2- and 3- substituted compounds being produced.

Subsequent metabolism of the alcohol may follow a number of pathways as illustrated in Fig. 1. Following terminal oxidation, the produced alcohol is normally oxidised to the corresponding aldehyde and fatty acid by means of pyridine nucleotide-linked dehydrogenases. In some *Candida* spp., alcohol oxidases have been shown to be present in place of alcohol dehydrogenases. This has been reported, for example, by Blasig et al. (1988) for *Candida maltosa*, Kemp et al. (1988) for *C. tropicalis* and Hommel & Ratledge (1990) for *C. bombicola*. As an alternative to monoterminal oxidation, ω-oxidation may occur resulting in the production of either or both α,ω-dioic acids and ω-hydroxy fatty acids. Rehm et al. (1983) and Blasig et al. (1988, 1989) have recently described the action of both mono- and diterminal oxidation systems in *Mortierella isabellina* and *Candida* spp., respectively. Woods & Murrell (1989) have reported that the propane-oxidising bacterium *Rhodococcus rhodochrous* can produce monoterminal and diterminal oxidation of C_2–C_8 n-alkanes *via* a system that is not linked to cytochrome P-450. In a survey of n-alkane oxidation in a number of microbial species, Rehm & Reiff (1982) observed diterminal oxidation products only rarely. All of these products may be further metabolized by means of the β-oxidation pathway for fatty acids. Subterminal alcohols are oxidised to the corresponding ester and hydrolytically cleaved to produce an acid and an alcohol. Following oxidation of the alcohol, the fatty acids produced may be metabolized via normal cellular pathways.

There is also some weak evidence for n-alkane metabolism via alkenes produced by the action of a NAD(P)-linked dehydrogenase. The alkene is purported to be hydroxylated across the double bond and further metabolized as described above. The existence of this pathway appears to be relatively dubious and even when reported is described as being relatively slow (Ratledge 1978; Singer & Finnerty 1984a). However, such a pathway does represent a means of alkane degradation independent of oxygenase activity and may therefore represent a potential route for anaerobic attack upon alkanes. This possibility is considered in more detail below. Recent observations of aromatic hydrocarbon degradation under anaerobic conditions also make one wary of discounting the possibility of other pathways for aliphatic hydrocarbon breakdown under anaerobic conditions.

Metabolism of branched-chain alkanes

Branched-chain alkanes tend to be less readily degraded than n-alkanes and in hydrocarbon mixtures degradation of branched compounds is generally repressed by the presence of straight-chain substrates (Pirnik et al. 1984). It is possible to make very general assertions about the relationship between the structure of branched alkanes and their degradability. Highly branched compounds are more recalcitrant to biodegradation than simpler compounds. Particularly recalcitrant are β-branched (anteiso-) and quaternary branched compounds due to steric hindrance of oxidation enzymes (Britton 1984). However, detailed correlation of structure and biodegradability is not possible (Singer & Finnerty 1984a) and even simple generalisations such as those given above do not hold true in all cases. For example, quaternary compound degradation has been described as in the case of the conversion of 2,2-dimethylheptane to 2,2-dimethylpropionate by '*Achromobacter*' sp. as cited by Singer & Finnerty (1984a).

It is becoming evident that degradation of a diverse range of branched alkanes can occur and that much of the reported recalcitrance of such compounds is due to the absence of suitable experimentation. The isoprenoid hydrocarbon pristane (2,6,10,14-tetramethylpentadecane) is commonly used as an internal marker in environmental hydrocarbon analysis since it is viewed as being highly persistent during the degradation of crude oil and petroleum products. However, its degradation has been widely studied and clearly elucidated in '*Brevibacterium*' sp. (Pirnik et al. 1974), *Corynebacterium* sp. (McKenna & Kallio 1971) and *Rhodococcus* sp. (Nakajima & Sato 1983). The metabolic pathways involved have been reviewed by Pirnik (1977) and degradation may occur by β- or ω-oxidation as illustrated in Fig. 2. Other complex branched alkanes have also been shown to be metabolized. Nakajima et al. (1985) described a *Rhodococcus* sp. capable of degrading phytane (2,6,10,14-tetramethylhexadecane), norpristane (2,6,10-trimethylpentadecane) and farnesane (2,6,10-trimethyldodecane) as sole sources of carbon and energy. In all cases isopropyl units on the molecules were oxidised to terminal alcohols and thence to the corresponding acids. Cox et al. (1976) reported a *Mycobacterium* sp. that could degrade phytane, norpristane, 2,6,10-trimethyltetradecane and 2,6,10,14-tetramethylheptadecane. Unlike the report of Nakajima & Sato (1983), initial attack upon the molecules did not occur only at isopropyl

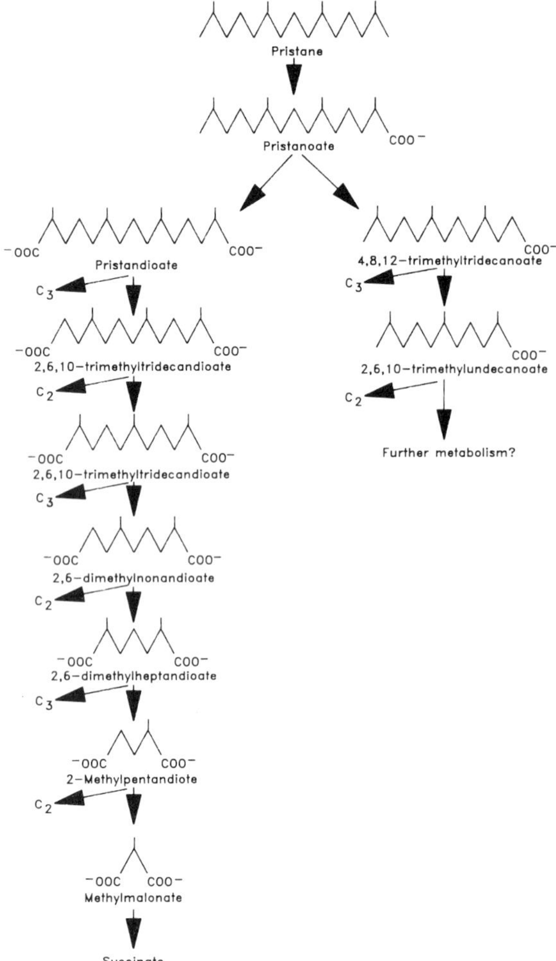

Fig. 2. Metabolic routes for pristane (2,6,10,14-tetramethylpentadecane) by microorganisms. This pathway serves as a model for the degradation of isoprenoid aliphatic hydrocarbons in bacteria. From Britton (1984), reprinted by courtesy of Marcel Dekker Inc.

termini but the products of initial oxidation were always terminal alcohols. In an exciting development, Rontani & Giusti (1986) have reported the utilisation of 2,2,4,4,6,8,8-heptamethylnonane as a sole carbon and energy source by a mixed marine microbial population. Degradation was relatively rapid and the only metabolic intermediates detected were straight-chain fatty acids. The metabolic pathway involved oxidation at the β-position to give the corresponding ketone. This was then oxidised to an ester which was hydrolysed prior to β-oxidation. A range of other highly-branched alkanes could be utilised provided that there was a free β-carbon available for oxidation.

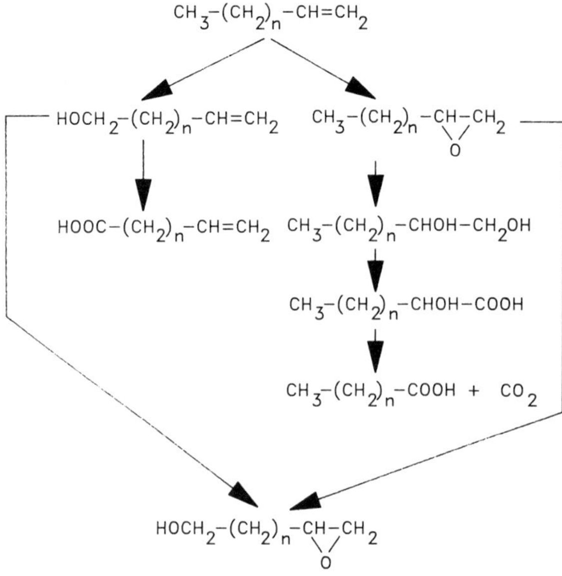

Fig. 3. Basic metabolic pathways for the degradation of alkenes in microorganisms. Initial attack on the molecule may be across the double bond to produce a subterminal epoxide or diol or may be analogous to n-alkane oxidation and leave the double bond initially intact.

Metabolism of alkenes

The metabolism of alkenes may be initiated either by attack upon the double bond or by an oxidation reaction elsewhere in the molecule as occurs for n-alkanes. Britton (1984) recognised four major patterns of initial attack: oxygenase attack upon a terminal methyl group to the corresponding unsaturated alcohols and acids; subterminal oxygenase attack to the corresponding alcohols and acids; oxidation across the double bond to the corresponding epoxide; oxidation across the double bond to the corresponding diol. A basic outline of the processes of initial metabolism is given in Fig. 3 but it is important to note that organisms may exhibit more than one of these pathways.

Recent studies have concentrated primarily on the metabolism of short-chain (hexene and below) alkenes and this work has been thoroughly reviewed by Hartmans et al. (1989). The basic pathways involved in the metabolism of short-chain compounds are as described above but several interesting observations have been made. van Ginkel & de Bont (1986) and van Ginkel et al. (1987) obtained *Mycobacterium, Nocardia* and *Xanthobacter* strains growing on short-chain alkenes which were unable to metabolize the corresponding alkanes since the only metabolic pathway present involved epoxidation of the double bond. This possession of a single pathway is very common in microorganisms capable of utilising short-chain alkenes but rare in those utilising longer-chain compounds. Particular interest has been focused upon the degradation of propene since this compound presents special difficulties to the degradative organisms. Oxidation reactions may result in the production of acrylate and then the non-degradable acrylate-CoA which thereby depletes the cellular CoA pool. As a means of overcoming this, strain PL-1 reported by Cerniglia et al. (1976) splits propene across the double bond into C_1 and C_2 units. The propene-utilising *Mycobacterium* and *Xanthobacter* strains described by de Bont et al. (1980) and van Ginkel and de Bont (1986) perform the same process *via* 1,2-epoxypropane.

Note that the aerobic degradation of alkynes can also occur, probably by means of hydratase activity (Hartmans et al. 1989).

Anaerobic degradation of aliphatic hydrocarbons

The anaerobic degradation of hydrocarbons has been a topic of intense scientific debate for many years. Recently, reliable evidence has been obtained demonstrating the degradation of aromatic hydrocarbons in anoxic environments (see later review by Smith). The anaerobic degradation of alkanes is a subject that still arouses diverse opinions. Bertrand et al. (1989) appear to be optimistic concerning the possibility of anaerobic alkane degradation despite the absence of convincing evidence that this is a real phenomenon. In contrast, Schink (1989) remains unconvinced of the possibility of anaerobic alkane degradation. Schink (1985a) has calculated that the anaerobic metabolism of alkanes to produce methane is an energetically favourable process but remains sceptical of the possibility that such processes occur since suitable enzyme systems to bring about such reactions have not been observed to date despite extensive research.

The most convincing evidence to date for the anaerobic degradation of aliphatic hydrocarbons has come from Schink (1985a). He added a large number of hydrocarbons to methanogenic enrichment cultures and monitored increased methane production due to their degradation. Of the substrates tested, including n-alkanes, branched alkanes and alkenes, only two, 1-hexadecene and squalene, increased methanogenesis. Repeated sub-culturing to ensure the complete elimination of alternative carbon sources confirmed the apparent degradation. For hexadecene, the extra methane detected accounted for 78–91% of the theoretical methane producible for this substrate. For squalene growth was much poorer and methanogenesis stopped after a period. It was calculated that the $\Delta G_0'$ values of such reactions were favourable for microbial growth and proposed that hydration across the double bond to produce an alcohol was the most likely reaction. However, Schink (1989) has noted that this may be either a biological or an abiotic chemical reaction and that there is no evidence for or against either of these possibilities. Thus, it can be concluded that attack upon alkenes appears feasible but substrate range has so far been found to be limited. For alkanes it would still appear that conclusive evidence is lacking. Certainly the potential metabolic reactions are energy-yielding but no enzymes have yet been described which are capable of initiating attack. It is possible that the proposed reduction of alkanes to alkenes could be a feasible route but this has not yet been conclusively demonstrated (Singer & Finnerty 1984a) and has certainly not been seen anaerobically.

One special case of anaerobic aliphatic hydrocarbon degradation is that of ethyne. This compound is commonly degraded anaerobically (Hartmans et al. 1989) and may also be converted by a variety of non-specific mechanisms, including that of nitrogenase. Degradation of ethyne has been obtained in pure culture. Schink (1985b) reported a new spe-

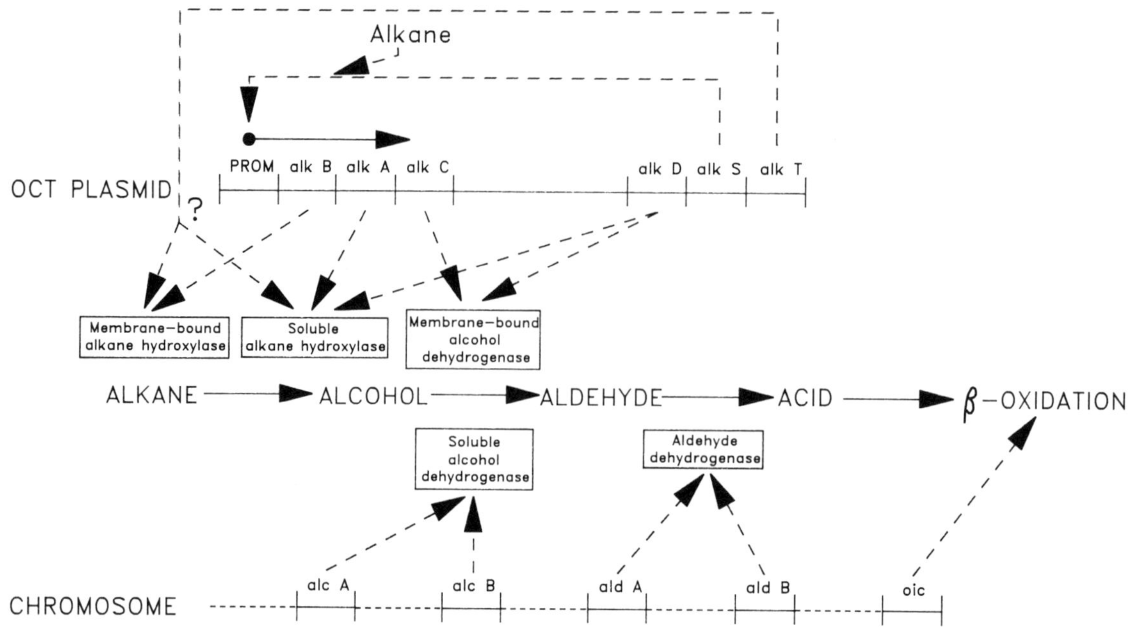

Fig. 4. Structure and function of the *OCT* plasmid and associated chromosomal genes in *Pseudomonas putida* PpG6. The *alk S* and *alk R* loci correspond to the area originally identified as the *alk R* regulatory locus. The *alk S* gene product is a regulatory protein that controls transcription of the *alk BAC* operon which is induced by n-alkanes. The role of the *alk T* gene product is unclear but it appears to be part of the alkane hydroxylase enzyme complexes. Chromosomal genes code for the soluble alcohol dehydrogenase, aldehyde dehydrogenase and β-oxidation enzymes. Information from Singer & Finnerty (1984a) and Eggink et al. (1987a, 1988).

cies (*Pelobacter acetylinicus*) that could grow on ethyne as a sole carbon and energy source. The degradation probably involved hydration to acetaldehyde followed by disproportionation to acetate and ethanol.

Genetics of aliphatic hydrocarbon degradation

The genetics of aliphatic hydrocarbon degrading organisms has only been studied in detail in a small number of microorganisms. The best studied system in bacteria is the *OCT* plasmid which codes for a number of proteins involved in growth on C_6 to C_{10} n-alkanes. This plasmid was originally intensively studied in *Pseudomonas putida* PpG6 (ATCC 17633) and the induction of the system, basic metabolic pathways and genetics were reported in a number of papers, most notably Chakrabarty et al. (1973), Nieder & Shapiro (1975), Grund et al. (1975) and Benson et al. (1977). The metabolic pathway for n-alkanes was found to involve an initial monooxygenation reaction by an alkane hydroxylase to produce the corresponding 1-ol. This was further oxidised to the corresponding aldehyde by an aldehyde dehydrogenase and then to the fatty acid by means of an aldehyde dehydrogenase. The fatty acids were then metabolized *via* the β-oxidation pathway. Fennewald and co-workers (Fennewald & Shapiro 1977, 1979; Fennewald et al. 1979) identified a number of loci on the plasmid and the chromosome of *Pseudomonas aeruginosa* which were involved in coding for the pathway. The designation of these was as follows: *alk* for alkane hydroxylase; *ald* for aldehyde dehydrogenase; *alc* for alcohol dehydrogenase; *oic* for a locus involved in the β-oxidation pathway. Genes identified on the chromosome were: *alc A* coding for soluble alcohol dehydrogenase with a substrate range of C_7 and above; *alc B* coding for soluble alcohol dehydrogenase with a substrate range of C_3 to C_6; *ald A* and *ald B* coding for the corresponding aldehyde dehydrogenases; *oic*. Among the apparent loci located on the plasmid were a promoter sequence, regulatory genes and genes coding for subunits of alkane hydroxylase and alcohol dehydrogenase.

Extensive further research has enabled the construction of a detailed picture of the operation of the *OCT* system as illustrated in Fig. 4. Eggink et al. (1987a) cloned a region of *OCT* DNA containing the promoter, the *alk BAC* operon and the regulatory locus (then known as *alk R*) into *Escherichia coli* and non-degradative strains of *Pseudomonas putida*. Expression of the genes occurred demonstrating that this fragment contained all of the necessary information for production of the alkane hydroxylase and alcohol dehydrogenase activity. Using a 16.9 kilobasepair *Eco* RI fragment, Eggink et al. (1987b) were able to translate the genes of the *alk BAC* operon in an *E. coli* minicell preparation. They detected six proteins, of which four could be assigned a definite function: 41 kDa, membrane-bound alkane dehydrogenase; 15 kDa and 49 kDa, subunits of soluble alkane hydroxylase; 58 kDa, probably involved in membrane-bound alcohol dehydrogenase activity; 59 kDa; 20 kDa. The 41 kDa protein was associated with the *alk B* locus, the soluble alkane hydroxylase subunits with the *alk A* locus and the *alk C* locus appeared to code for the 58 kDa polypeptide. Kok et al. (1989) further studied the *alk B* region by sequencing. They defined a 401 amino-acid polypeptide with 8 hydrophobic regions of suitable size for spanning cell membranes. Extensive investigation of the *alk R* region suggested that there may be several operons present at this location (Owen 1986). Eggink et al. (1988) translated a 4.9 kilobasepair fragment from this region in an *E. coli* minicell preparation and identified two cistrons designated *alk S* and *alk T*. The former coded for a 99 kDa polypeptide which directly promoted transcription of the *alk BAC* operon. The *alk T* gene product was a 48 kDa protein of uncertain function although it was necessary for activity of alkane hydroxylase and did not act as a regulator of gene expression. It was concluded that this protein was part of the alkane hydroxylase complex.

Genetics of other alkane-oxidising bacteria has been far less extensively studied. Singer & Finnerty (1984b) reviewed observations on the genetics of alkane degradation in *Acinetobacter* HO1-N which can only utilise alkanes with a carbon chain length of C_{10} or above. Despite an extensive search no plasmids coding for any regions of the degradative pathway have been detected and it has been concluded that the pathway is encoded on the chromosome. Unlike the *OCT* system, only one set of enzymes is involved in oxidation of compounds of all chain lengths. Analysis of non-degradative mutants has demonstrated that the region of the chromosome coding for the alcohol and aldehyde dehydrogenase enzymes is totally independent of that coding for the initial alkane-oxidising enzyme.

The genetics of hydrocarbon degradation in yeasts has been little studied. Bassel & Mortimer (1985) illustrated the complexity of the genetic system in yeasts by identifying at least 16 loci that control functions affecting n-alkane uptake in *Yarrowia lipolytica*. One particular area that is receiving increasing interest has been the alkane-inducible cytochrome P-450 detected in certain *Candida* species. In *C. tropicalis* sequence analysis has shown that this inducible cytochrome is distinctly different from other P-450 cytochromes previously reported and it has been concluded that this represents a member of a new family of cytochrome P-450 (Sanglard et al. 1987; Sanglard & Loper 1989). A second cytochrome of this family has now been detected in alkane-induced *C. tropicalis*. The family has been termed P-450 LII and the cytochromes designated P-450 *alk 1* and *alk 2* (Sanglard & Fiechter 1989). A very similar inducible cytochrome P-450 has also been observed in *C. maltosa* (Suniari et al. 1988). Sequencing of this has shown that it is very closely related to the LII gene family of *C. tropicalis* and has shown that there are two highly hydrophobic regions of the molecule towards the N-terminus which apparently serve to anchor the molecule in the endoplasmic reticulum (Takagi et al. 1989; Schunk et al. 1989).

Biotechnological applications of aliphatic hydrocarbon metabolism

Aliphatic hydrocarbon-degrading microorganisms may have a number of potential biotechnological

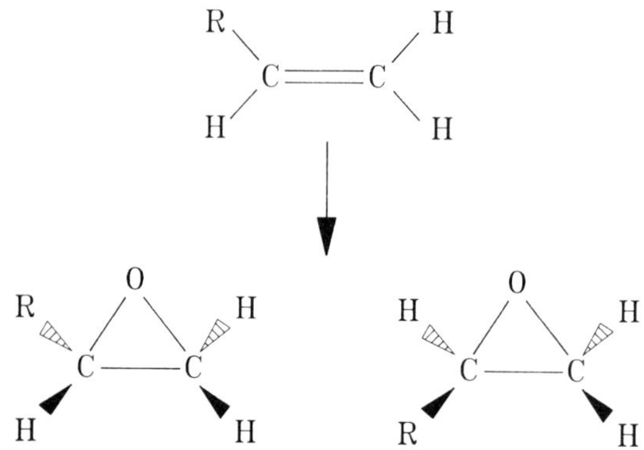

Fig. 5. Illustration of production of stereoisomers during epoxidation across the double bond of alkenes with a chain length of three carbon atoms or more.

applications. Much of the early research on biotechnology based upon these organisms focused upon the use of hydrocarbon substrates as feedstocks for single cell protein production but this proved to be an uneconomical process following the oil price rises of the early 1970s. More recent research has highlighted three areas where the physiology of aliphatic hydrocarbon degradation may have possible biotechnological application and these will be considered in turn.

Epoxide production

Epoxides are widely used as feedstocks in the synthetic chemical industry. The production of epoxides from alkenes has been extensively documented. When produced by co-oxidation reactions (e.g. by methane monooxygenase; MMO), by some alkane oxidation enzymes or by mutant strains blocked in the subsequent stage of alkene degradation, they may accumulate in the culture medium (Hartmans et al. 1989). Although epoxides are relatively toxic to microorganisms, developments in strain selection and bioengineering mean that it is possible to utilise bioreactor systems for the synthesis of epoxides. In particular, biocatalytic conversion of hydrocarbon substrates in two-phase culture systems have been shown to be widely applicable (de Smet et al. 1983a; Witholt et al. 1990). For the extensively used compounds epoxyethane and epoxypropane it is unlikely that biotechnologies will be able to compete on cost with chemical syntheses. However, alkenes of three carbon atoms or more are asymmetric when converted to epoxides and this permits the production of stereoisomers (Fig. 5). Chemical syntheses and some microbiological enzymes produce racemic mixtures but some microorganisms produce epoxides which are virtually optically pure (Weijers et al. 1988a). There is extensive interest in the use of specific stereoisomers as substrates for chemical syntheses in the pharmaceutical and fine chemicals industries and the potential application of biotechnology for stereospecific syntheses has been widely considered (Hartmans et al. 1989; Weijers et al. 1988b,c).

Production of aliphatic hydrocarbon-based metabolites

It has long been known that growth on specific hydrocarbon substrates results in the enrichment of specific fatty acids in the cell envelope (Ratledge 1978; Finnerty 1984). This enrichment could conceivably have a biotechnological role if the enriched compounds included a high-value fatty acid that was difficult to obtain from other sources, but is otherwise unlikely to be economically viable as a technology. The same would appear to be true of other compounds that can be obtained readily from other sources. For example, Shennan (1984) and Miall (1980) review the efforts devoted to the biotechnological production of citric acid and amino-acids by alkane-degrading microorganisms which were subsequently rendered uneconomical. Other metabolites may be of greater interest. Biosurfactants produced during growth on hydrocarbons have received a good deal of attention and their potential is reviewed elsewhere in this volume by Hommel. Ratledge (1984) reviewed work on the composition of wax esters produced by microorganisms during growth on alkanes. The composition and degree of unsaturation of these esters varies according to the substrate provided and certain waxes may represent viable alternatives to materials currently derived from plant or animal sources. Extensive further work is necessary to develop this technology towards actual deployment.

De Smet et al. (1983b) reported the production of a poly-3-hydroxyalkanoate intracellular storage heteropolymer during the growth of *Pseudomonas oleovorans* on n-octane. The composition and degree of unsaturation of this polymer can be varied by cultivation of this organisms on C_6 to C_{12} n-alkanes and C_8 to C_{10} n-alkenes (Lageveen et al. 1988). By selection of the substrate and growth conditions, yields of up to approximately 25 g polymer per 100 g cell dry weight can be obtained in a fermentation process lasting 20 to 30 hours. These compounds have potential application in the production of biologically-derived plastics.

Environmental biotechnology

The use of biodegradative microorganisms to clean oil-contaminated soil and water is receiving increasing attention (Morgan & Watkinson 1989a). Exploitation of the physiology of aliphatic hydrocarbon-degrading organisms will of necessity be part of this process, but at the majority of locations it is probable that the indigenous microbial population will be adapted for degradation of the aliphatic hydrocarbons present. However, in special circumstances, when the indigenous population is prevented from developing by local environmental conditions or when particular substrates are to be degraded, it may be necessary to inoculate contaminated locations with specific degradative phenotypes. The co-oxidation of aliphatic hydrocarbons by methylotrophic bacteria has potential applications in environmental cleanup (Morgan & Watkinson 1989b) and may also be useful in biofilters for the cleaning of hydrocarbon-containing waste gas streams (Hartmans et al. 1989). Further research into these areas is necessary.

Conclusions

The microbial degradation of aliphatic hydrocarbons has long been known and many of the basic features of the physiology of degradative organisms had been considered to be well documented. Emphasis in recent years has tended to move away from studying aliphatic hydrocarbon degradation but recent developments in a number of key areas have shown that there is much still to be studied, for example pathways for the degradation of branched alkanes. The genetics and regulation of the degradation pathways for aliphatic hydrocarbons has recently made major advances, particularly in our understandings of the structure and function of the genes in the *OCT* plasmid. However, our knowledge of the pathways in other organisms has been little studied. Also requiring further elucidation are the detailed molecular processes involved in hydrocarbon transport into microbial cells. The potential for anaerobic degradation of most aliphatic hydrocarbons remains unclear and raises a number of fundamental scientific questions about the physiology of hydrocarbon degradation. Finally, attempts are now being made to exploit the physiological activity of alkane and alkene-metabolizing microorganisms in a number of areas of biotechnology. It is evident that much research is still necessary in these and other fields. It is hoped that future results will continue to be as interesting and novel as those discussed in this article.

References

Atlas RM (1981) Microbial degradation of petroleum hydrocarbons: an environmental perspective. Microbiol Rev 45: 180–209

Bassel JB & Mortimer RK (1985) Identification of mutations preventing n-hexadecane uptake among 26 n-alkane non-utilizing mutants of *Yarrowia (Saccharomycopsis) lipolytica*. Curr Genet 9: 579–586

Benson S, Fennewald M, Shapiro J & Huettner C (1977) Fractionation of inducible alkane hydroxylase activity in *Pseudomonas putida* and characterization of hydroxylase-negative plasmid mutations. J Bacteriol 132: 614–621

Bertrand JC, Caumette P, Mille G, Gilewicz M & Denis M (1989) Anaerobic biodegradation of hydrocarbons. Sci Prog 73: 333–350

Blasig R, Mauersberger S, Riege P, Schunck W-H, Jockisch W, Franke P & Muller H-G (1988) Degradation of long-chain n-alkanes by the yeast *Candida maltosa*. II. Oxidation of n-alkanes and intermediates using microsomal membrane fractions. Appl Microbiol Biotechnol 28: 589–597

Blasig R, Huth J, Franke P, Borneleit P, Schunck, W-H & Muller H-G (1989) Degradation of long-chain n-alkanes by the yeast *Candida maltosa*. III. Effect of solid n-alkanes on cellular fatty acid composition. Appl Microbiol Biotechnol 31: 571–576

de Bont JAM, Primrose SB, Collins MD & Jones D (1980) Chemical studies on some bacteria which utilise gaseous unsaturated hydrocarbons. J Gen Microbiol 117: 97–102

Britton LN (1984) Microbial degradation of aliphatic hydrocarbons. In: Gibson DT (Ed) Microbial Degradation of Organic Compounds (pp 89-129). Marcel Dekker, New York

Cerniglia CE, Blevins WT & Perry JJ (1976) Microbial oxidation and assimilation of propylene. Appl Environ Microbiol 32: 764–768

Chakrabarty AM, Chou G & Gunsalus IC (1973) Genetic regulation of octane dissimilation plasmid in *Pseudomonas*. Proc Natl Acad Sci USA 70: 1137–1140

Cox RE, Maxwell JR & Myers RN (1976) Monocarboxylic acids from oxidation of acyclic isoprenoid alkanes by *Mycobacterium fortuitum*. Lipids 11: 72–76

Eastcott L, Shiu WY & Mackay D (1988) Environmentally relevant physical-chemical properties of hydrocarbons: a review of data and development of simple correlations. Oil Chem Pollut 4: 191–216

Eggink G, Engel H, Meijer WG, Otten J, Kingma J & Witholt B (1988) Alkane utilization in *Pseudomonas oleovorans*. Structure and function of the regulatory locus alkR. J Biol Chem 263: 13400–13405

Eggink G, Lageveen RG, Altenburg B & Witholt B (1987a) Controlled and functional expression of the *Pseudomonas oleovorans* alkane utilizing system in *Pseudomonas putida* and *Escherichia coli*. J Biol Chem 262: 17712–17718

Eggink G, van Lelyveld PH, Arnberg A, Arfman N, Witteveen C & Witholt B (1987b) Structure of the *Pseudomonas putida* alkBAC operon. Identification of transcription and translation products. J Biol Chem 262: 6400–6406

Fennewald M, Benson S, Oppici M & Shapiro J (1979) Insertion element analysis and mapping of the *Pseudomonas* plasmid alk region. J Bacteriol 139: 940–952

Fennewald M & Shapiro J (1977) Regulatory mutations of the *Pseudomonas* plasmid alk regulon. J Bacteriol 132: 622–627

Fennewald M & Shapiro J (1979) Transposition of Tn7 in *Pseuomonas aeruginosa* and isolation of alk : : Tn7 mutations. J Bacteriol 139: 264–269

Finnerty WR (1984) The application of hydrocarbon-utilizing microorganisms for lipid production. AOCS Monogr 11: 199–215

Finnerty WR & Singer ME (1985) Membranes of hydrocarbon-utilizing microorganisms. In: Ghosh BK (Ed) Organisation of Prokaryotic Cell Membranes, Volume III (pp 1–44). CRC Press, Boca Raton, Florida

Fukui S & Tanaka A (1979) Peroxisimes of alkane- and methanol-grown yeasts: metabolic functions and practical applications. J Appl Biochem 1: 171–201

van Ginkel CG & de Bont JAM (1986) Isolation and characterization of alkene-utilizing *Xanthobacter* spp. Arch Microbiol 145: 403–407

van Ginkel CG, Welten HGJ & de Bont JAM (1987) Oxidation of gaseous and volatile hydrocarbons by selected alkene-utilizing bacteria. Appl Environ Microbiol: 2903–2907

Grund A, Shapiro J, Fennewald M, Bacha P, Leahy J, Markbreiter K, Nieder M & Toepfer M (1975) Regulation of alkane oxidation in *Pseudomonas putida*. J Bacteriol 123: 546–556

Hartmans S, de Bont JAM & Harder W (1989) Microbial metabolism of short-chain unsaturated hydrocarbons. FEMS Microbiol Rev 63: 235–264

Hommel R & Kleber H-P (1984) Oxidation of long-chain alkanes by *Acetobacter rancens*. Appl Microbiol Biotechnol 19: 110–113

Hommel R & Ratledge C (1990) Evidence for two fatty alcohol oxidases in the biosurfactant-producing yeast *Candida (Torulopsis) bombicola*. FEMS Microbiol Lett 70: 183–186

Kappeli O & Finnerty WR (1979) Partition of alkane by an extracellular vesicle derived from hexadecane-grown *Acinetobacter*. J Bacteriol 140: 707–712

Kemp GD, Dickinson FM & Ratledge C (1988) Inducible long chain alcohol oxidase from alkane-grown *Candida tropicalis*. Appl Microbiol Biotechnol 29: 370–374

Kennicutt MC (1988) The effect of biodegradation on crude oil bulk and molecular composition. Oil Chem Pollut 4: 89–112

Kirk PW & Gordon AS (1988) Hydrocarbon degradation by filamentous marine higher fungi. Mycologia 80: 776–782

Kok M, Oldenhuis R, van der Linden MPG, Raatjes P, Kingma J, van Lelyveld PH & Witholt B (1989) The *Pseudomonas oleovorans* alkane hydroxylase gene. Sequence and expression. J Biol Chem 264: 5435–5441

Lageveen RG, Huisman GW, Preusting H, Ketelaar P, Eggink G & Witholt B (1988) Formation of polyesters by *Pseudomonas oleovorans*: effect of substrates on formation and composition of poly-(R)-3-hydroxyalkanoates and poly-(R)-3-hydroxyalkenoates. Appl Environ Microbiol 54: 2924–2932

Lindley ND, Pedley JF, Kay SP & Heydeman M.T. (1986) The metabolism of yeasts and filamentous fungi which degrade hydrocarbon fuels. Int Biodet 22: 281–287

McKenna EJ & Kallio RE (1971) Microbial metabolism of the isoprenoid alkane pristane. Proc Natl Acad Sci USA 68: 1552-1554

Miall LM (1980) Organic acid production from hydrocarbons. In: Harrison DEF, Higgins IJ & Watkinson RJ (Eds) Hydrocarbons in Biotechnology (pp 25-34). Heyden, London

Mille G, Mulyono, M, El Jammel T & Bertrand J-C (1988) Effects of oxygen on hydrocarbon degradation studies in vitro in surficial sediments. Estuarine Coastal Shelf Sci 27: 283–295

Miller RM & Bartha R (1989) Evidence for liposome encapsulation for transport-limited microbial metabolism of solid alkanes. Appl Environ Microbiol 55: 269–274

Morgan P & Watkinson RJ (1989a) Hydrocarbon degradation in soils and methods for soil biotreatment. CRC Crit Rev Biotechnol 8: 305–333

Morgan P & Watkinson RJ (1989b) Microbiological methods

for the cleanup of soil and ground water contaminated with halogenated organic compounds. FEMS Microbiol Rev 63: 277–300

Nakajima K & Sato A (1983) Microbial metabolism of isoprenoid alkane pristane. Nippon Nogeikugaku Kaishi 57: 299–305

Nakajima K, Sato A, Takahara Y & Iida T (1985) Microbial oxidation of isoprenoid alkanes, phytane, norpristane and farnesane. Agric Biol Chem 49: 1493–2002

Nakajima K, Sato A, Takahara Y & Iida T (1985) Microbial oxidation of isoprenoid alkanes phytane, norpristane and farnesane. Agric Biol Chem 49: 1993–2002

Ng TK & Hu WS (1989) Adherence of emulsan-producing *Acinetobacter calcoaceticus* to hydrophobic liquids. Appl Microbiol Biotechnol 31: 480–485

Nieder M & Shapiro J (1975) Physiological function of the *Pseudomonas putida* PpG6 (*Pseudomonas oleovorans*) alkane hydroxylase: monoterminal oxidation of alkanes and fatty acids. J Bacteriol 122: 93–98

Oudot J, Ambles A, Bourgeouis S, Gatellier C & Sebyera N (1989) Hydrocarbon infiltration and biodegradation in a landfarming experiment. Environ Pollut 59: 17–40

Owen DJ (1986) Molecular cloning and characterization of sequences from the regulatory cluster of the *Pseudomonas* plasmid alk system. Mol Gen Genet 203: 64–72

Pfaender FK & Buckley EN (1984) Effects of petroleum on microbial communities. In: Atlas RM (Ed) Petroleum Microbiology (pp 507–536). Macmillan, Mew York

Pirnik M.P. (1977) Microbial oxidation of methyl branched alkanes. CRC Crit Rev Microbiol 5: 413–422

Pirnik MP, Atlas RM & Bartha R (1974) Hydrocarbon metabolism by *Brevibacterium erythrogenes*: normal and branched alkanes. J Bacteriol 119: 868–878

Ratledge C (1978) Degradation of aliphatic hydrocarbons. In: Watkinson RJ (Ed) Developments in Biodegradation of Hydrocarbons (pp 1–46). Applied Science, London

Ratledge C (1984) Microbial conversions of alkanes and fatty acids. J Am Oil Chem Soc 61: 447–453

Rehm HJ, Hortmann L & Reiff I (1983) Regulation der fettsaurebildung bei der mikrobiellen alkanoxidation. Acta Biotechnol 3: 279–288

Rehm HJ & Reiff I (1982) Regulation der mikrobiellen alkanoxidation mit hinblick auf die produktbildung. Acta Biotechnol 2: 127–138

Rontani JF & Giusti G (1986) Study of the biodegradation of poly-branched alkanes by a marine bacterial community. Mar Chem 20: 197–205

Sanglard D, Chen C & Loper JC (1987) Isolation of the alkane inducible cytochrome P450 (P450alk) gene from the yeast *Candida tropicalis*. Biochem Biophys Res Comm 144: 251–257

Sanglard D & Fiechter A (1989) Heterogeneity within the alkane-inducible cytochrome P450 gene family of the yeast *Candida tropicalis*. FEBS Lett 256: 128–134

Sanglard D & Loper JC (1989) Characterization of the alkane-inducible cytochrome P450 (P450alk) gene from the yeast *Candida tropicalis*: identification of a new P450 gene family. Gene 76:121–136

Schink B (1985a) Degradation of unsaturated hydrocarbons by methanogenic enrichment cultures. FEMS Microbiol Ecol 31: 69–77

Schink B (1985b) Fermentation of acetylene by an obligate anaerobe *Pelobacter acetylenicus* sp. nov.. Arch Microbiol 142: 295–301

Schink B (1989) Anaerober abbau von Kohlenwasserstoffen. Erdol Kohle Erdgas 42: 116–118

Schunck W-H, Kargel E, Gross B, Wiedmann B, Mauersberger S, Kopke K, Kiessling U, Strauss M, Gaestel M & Muller H-G (1989) Molecular cloning and characterization of the primary structure of the alkane hydroxylating cytochrome P-450 from the yeast *Candida maltosa*. Biochem Biophys Res Comm 161: 843–850

Scott CCL & Finnerty WR (1976a) A comparative analysis of the ultrastructure of hydrocarbon-oxidizing microorganisms. J Gen Microbiol 94: 342–350

Scott CCL & Finnerty WR (1976b) Characterization of intracytoplasmic hydrocarbon inclusions from the hydrocarbon-oxidizing *Acinetobacter* sp. HO1-N. J Bacteriol 127: 481–489

Shennan JL (1984) Hydrocarbons as substrates in industrial fermentations. In: Atlas RM (Ed) Petroleum Microbiology (pp 643–683). Macmillan, New York

Singer ME & Finnerty WR (1984a) Microbial metabolism of straight-chain and branched alkanes. In: Atlas RM (Ed) Petroleum Microbiology (pp 1–59). Macmillan, New York

Singer ME & Finnerty WR (1984b) Genetics of hydrocarbon-utilizing microorganisms. In: Atlas RM (Ed) Petroleum Microbiology (pp 299–354). Macmillan, New York

de Smet M-J, Kingma J, Wijnberg H & Witholt B (1983a) *Pseudomonas oleovorans* as a tool in bioconversions of hydrocarbons: growth, morphology and conversion characteristics in different two-phase systems. Enzyme Microb Technol 5: 352–360

de Smet M-J, Eggink G, Witholt B, Kingma J & Wijnberg H (1983b) Characterization of intracellular inclusions formed by *Pseudomonas oleovorans* during growth on octane. J Bacteriol 154: 870–878

Sunairi M, Suzuki R, Takagi M & Yano K (1988) Self-cloning of genes for n-alkane assimilation from *Candida maltosa*. Agric Biol Chem 52: 577–579

Takagi M, Ohkuma M, Kobayashi N, Watanabe M & Yano J (1989) Purification of cytochrome P-450alk from n-alkane-grown cells of *Candida maltosa* and cloning and nucleotide sequencing of the encoding gene. Agric Biol Chem 53: 2217–2226

de Vries GM, Kues U & Stahl U (1990) Physiology and genetics of methylotrophic bacteria. FEMS Microbiol Rev 75: 57–101

Wakeham SG, Canuel EA & Doering PH (1986) Behavior of aliphatic hydrocarbons in coastal seawater: mesocosm experiments with [14C]octadecane and [14C]decane. Environ Sci Technol 20: 574–580

Watkinson RJ (1980) Interaction of microorganisms with hydrocarbons. In: Harrison DEF, Higgins IJ & Watkinson RJ (Eds) Hydrocarbons in Biotechnology (pp 11–24). Heyden, London

Weijers CGAM, van Ginkel CG & de Bont JAM (1988a) Enantiomeric composition of lower epoxyalkanes produced by methane-, alkane-, and alkene-utilizing bacteria. Enzyme Microb Technol 10: 214–218

Weijers CGAM, Leenen EJTM, Klijn N & de Bont JAM (1988b) Microbial formation of chiral epoxyalkanes. Med Fac Landbouww Rijksuniv Gent 53: 2089–2095

Weijers, CGAM, de Haan A & de Bont JAM (1988c) Microbial production and metabolism of epoxides. Microbiol Sci 5: 156–159

Witholt B, de Smet M-J, Kingma J, van Beilen JB, Kok M, Lageveen RG & Eggink G (1990) Bioconversions of aliphatic hydrocarbons by *Pseudomonas oleovorans* in multiphase bioreactors: background and economic potential. Trends Biotechnl 8: 46–52

Woods NR & Murrell JC (1989) The metabolism of propane in *Rhodococcus rhodochrous* PNKb1. J Gen Microbiol 135: 2335–2344

Biodegradation **1**: 93–105, 1990.

Microbial metabolism of monoterpenes – recent developments

Peter W. Trudgill
Department of Biochemistry, University College of Wales, Aberystwyth, Dyfed SY23 3DD, UK

Key words: 1,8-cineole, metabolism, monoterpenes, oxygenases, α-pinene, ring cleavage

Abstract

Monoterpenes are important renewable resources for the perfume and flavour industry but the pathways and enzymology of their degradation by microorganisms are not well documented. Until recently the acyclic monoterpene alcohols, (+)-camphor and the isomers of limonene were the only compounds for which significant sections of catabolic pathways and associated enzymology had been reported. In this paper recent developments in our understanding of the enzymology of ring cleavage by microorganisms capable of growth with 1,8-cineole and α-pinene are described. 1,8-Cineole has the carbocyclic skeleton of a monocyclic monoterpene with the added complication of an internal ether linkage. Ring hydroxylation strategy and biological Baeyer-Villiger oxygenation lead to an efficient method for cleaving the ether linkage. α-Pinene is an unsaturated bicyclic monoterpene hydrocarbon. At least two catabolic pathways exist. Information concerning one of them, in which α-pinene may be initially converted into limonene, is rudimentary. The other involves attack at the double bond resulting in formation of α-pinene epoxide. Ring cleavage is then catalysed by a novel lyase that requires no additional components and breaks both carbocyclic rings in a concerted manner.

Introduction

The monoterpenoids (C_{10}) are major components of plant oils and are synthesized from two isoprene units. Parent structures are acyclic, monocyclic or bicyclic and the latter consist of fused C6/C5, C6/C4 and C6/C3 ring systems. In addition to the parent hydrocarbons a number of oxygenated derivatives are also formed. Some representative structures are shown in Fig. 1.

Some of the earliest reported studies of monoterpene metabolism, by organisms capable of growth on them as sole carbon sources, were those of Seubert (1960), Seubert & Remberger (1963), Seubert et al. (1963), and Seubert & Fass (1964) using *Pseudomonas cintronellolis*. They are of particular interest since they firmly established a novel catabolic route for the degradation of cintronellol, geraniol and nerol in which oxidation of the primary alcohol to carboxyl and formation of a common CoA ester is followed by biotin-dependent carboxylation of a methyl group. Double bond hydration forms a 3-hydroxyacid CoA ester from which the two carbon unit side chain is eliminated as acetate by the action of a lyase. Two unimpeded cycles of β-oxydation are followed by a repetition of the methyl group elimination reactions which completes the degradation of these compounds to central metabolites (Fig. 2). Although this work was published over twenty five years ago the proposed pathway was supported by subcellular studies, enzyme isolation and ^{14}C experiments and, more recently, it has been extended to cover *P. putida* and *P. mendocia* strains by Cantwell et al. (1978). At the same time a group at the University of Illinois was investigating the degradation of the C6/C5 bicyclic monoterpene (+)-camphor by strains of *Pseudomonas putida* (Bradshaw et al.

CH2OH
Nerol
Limonene
1,8-Cineole
1,4-Cineole
Car-3-ene
α-Pinene
β-Pinene
(+)-Camphor

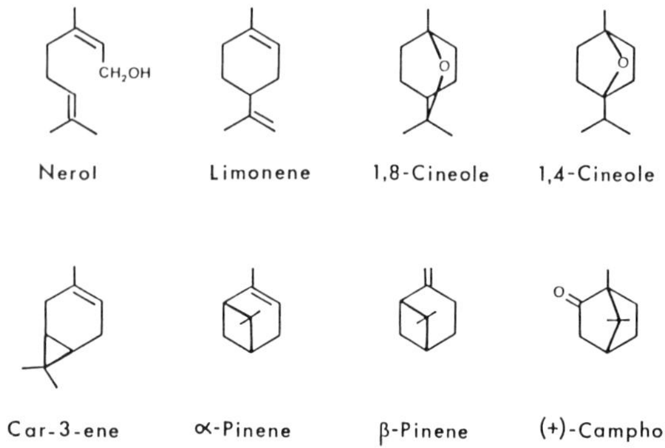

Fig. 1. Some naturally occurring monoterpenes.

1959; Conrad et al. 1961, 1965; Trudgill et al. 1966a, b) and *Mycobacterium rhodochrous* (Chapman et al. 1966). Although only limited success was achieved in understanding the catabolic pathways, key roles for methylene group hydroxylation and biological Baeyer-Villiger monooxygenases in ring cleavage strategies were established (Figs 3 and 4). No further experimental studies with *M. rhodochrous* T1 have been published but the (+)-camphor 5-hydoxylase from *P. putida* ATCC 17453 has been the subject of extensive and detailed research (see Gunsalus & Marshall 1971; Gunsalus & Lipscomb 1973; Gunsalus et al. 1974 for reviews). More recently, the oxygenating component of 2,5-diketocamphane 1,2-monooxygenase has been purified to homogeneity from *P. putida* (Taylor & Trudgill 1986) and the enzymology of the second ring cleavage step which involves CoA ester formation prior to biological Baeyer-Villiger oxygenation has been elucidated (Ougham et al. 1983; Trudgill 1986).

These particular areas of research probably represent the most detailed metabolic studies to date of monoterpene catabolism by organisms capable of growth with them as sole carbon sources although Bhattacharyya and his research group investigated the transformation of a number of monterpenes including geraniol, linalool, limonene and α-pinene and β-pinene, primarily by *Pseudomonas* strains. Very thorough and detailed analyses of metabolites accumulated in culture media were made. However, primarily because of the large number of metabolites accumulated, it was often difficult to identify a clear cut route leading to

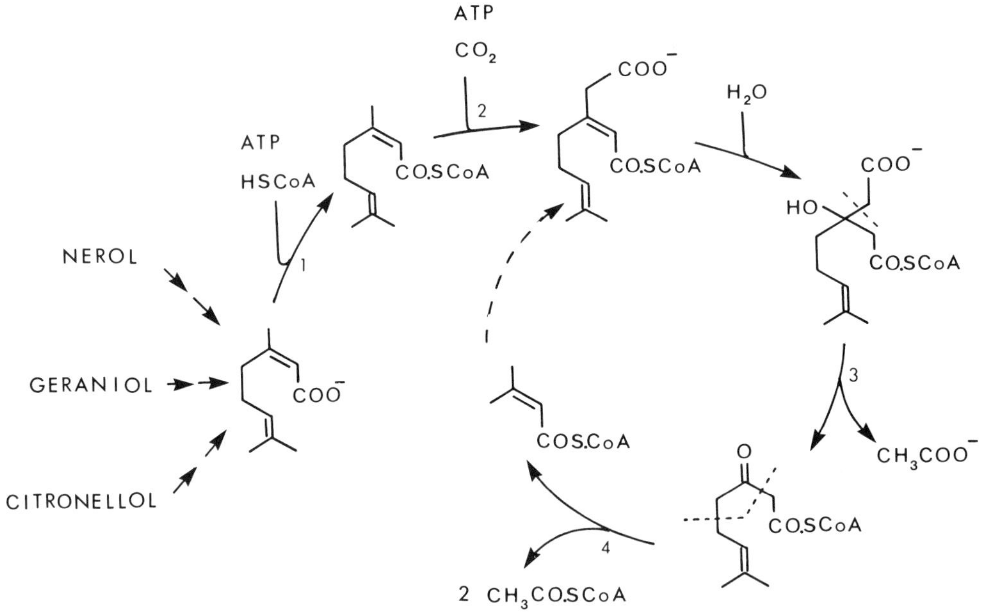

ATP
CO2
2
COO-
H2O
ATP
HSCoA
CO.SCoA
CO.SCoA
COO-
HO
CO.SCoA
NEROL
1
GERANIOL
COO-
3
COS.CoA
CH3COO-
CITRONELLOL
O
CO.SCoA
4
2 CH3CO.SCoA

Fig. 2. Degradation of acyclic monoterpene alcohols by *Pseudomonas citronellolis.* (*1*) CoA ester synthetase; (*2*) biotin-dependent carboxylase; (*3*) lyase; (*4*) two cycles of β-oxidation. From the work of Seubert (1960); Seubert & Remberger (1963); Seubert, Fass & Remberger (1963).

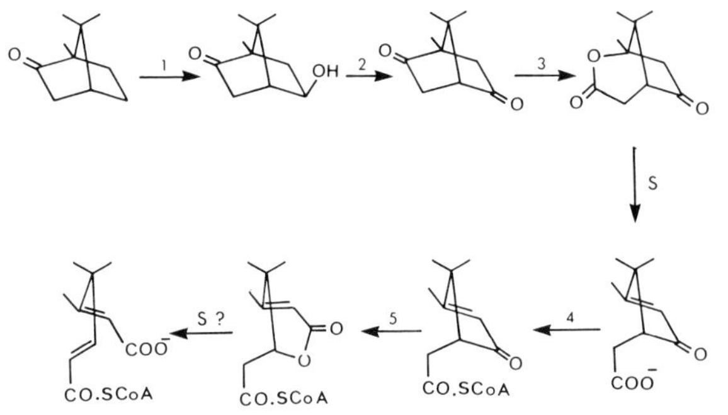

Fig. 3. Established steps of (+)-camphor degradation by *Pseudomonas putida* ATCC 17453. Enzymes: (*1*) camphor 5-hydroxylase; (*2*) 5-*exo*-hydroxy-camphor dehydrogenase; (*3*) 2,5-diketocamphane 1,2-monooxygenase (a Baeyer-Villiger monooxygenase); (*4*) 2-oxo-Δ^3-4,5,5-trimethylcyclopentenylacetyl-CoA synthetase; (*5*) 2-oxo-Δ^3-trimethylcyclopentenylacetyl-CoA monooxygenase (a Baeyer-Villiger monooxygenase). S = spontaneous reaction.

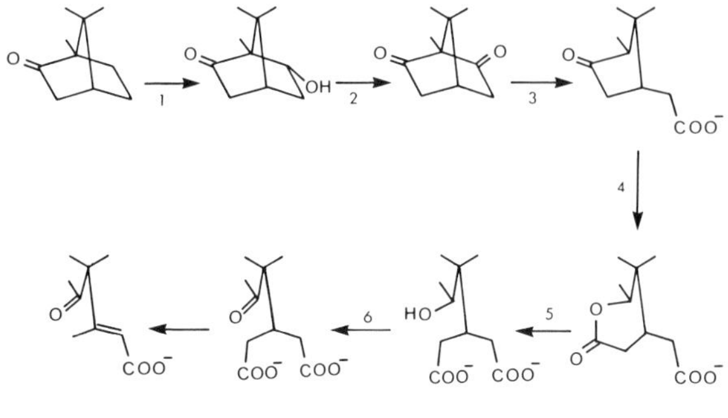

Fig. 4. Proposed steps of (+)-camphor degradation by *Mycobacterium rhodochorus* T1. Enzymes: (*1*) (+)-camphor 6-*endo*-hydroxylase; (*2*) 6-*end*-hydroxycamphor dehydrogenase; (*3*) 2,6-diketocamphane lyase; (*4*) 3-oxo-4,5,5-trimethylcyclopentanylacetic acid monooxygenase (a Baeyer-Villiger monooxygenase); (*5*) a lactone hydrolase; (*6*) a dehydrogenase. Based on results of Chapman, Meerman, Gunsalus, Srinivasan & Rinehart (1966).

central metabolites. Exceptions to this were the catabolism of limonene through perillic acid and a β-oxidative ring cleavage cycle (Dhavalikar & Bhattacharyya 1966; Dhavalikar et al. 1966) and evidence for the degradation of α-pinene through limonene or a closely related compound (Shukla & Bhattacharyya 1968; Shukla et al. 1968).

Current studies

More recent studies have been concentrated on two compounds where knowledge is limited and microorganisms are presented with particular structural complications in the growth substrate. The first of these is 1,8-cineole where the monocyclic monoterpene ring system has an ether linkage superimposed upon it and the second is α-pinene where work by Best et al. (1987) and ourselves (see below) has revealed a unique and quite novel ring cleavage enzyme.

1,8-Cineole

1,8-Cineole [1,3,3-trimethyl-2-oxabicyclo(2,2,2) octane] is isomeric with α-terpineol but contains neither a hydroxyl group nor a double bond. The trivial name eucalyptol is therefore rather misleading. It is found in considerable quantities in eucalyptus oil from the leaves of *Eucalyptus radiata* var Australiana (Nishimura et al. 1982). The molecule has the carbon skeleton of the monocyclic monoterpenes but, because of the ether linkage, is a bicyclic compound and it presents organisms capable of growing with it as sole carbon source with the additional problem of cleaving the ether linkage.

Metabolism of 1,8-cineole by Pseudomonas flava

Metabolite accumulation

Until recently, reports of microbial 1,8-cineole metabolism were confined to the isolation of metabolites from culture media and, following investigations of the dynamics of their production and disappearance, suggestions of a metabolic progression (MacRae et al. 1979; Carman et al. 1986). *Pseudomonas flava*, isolated by enrichment culture on 1,8-cineole accumulated the metabolites shown in Fig. 5. It is reasonable to suggest that initial attack occurs by hydroxylation at carbon 2, followed by dehydrogenation to form 2-oxocineole. Further metabolic steps are obscure, although a biological Baeyer-Villiger oxygenation can be envisaged. The source of the lactone (R)-5,5-dimethyl-4-(3′-oxobutyl)-4,5-dihydrofuran-2(3H)-one was not immediately evident.

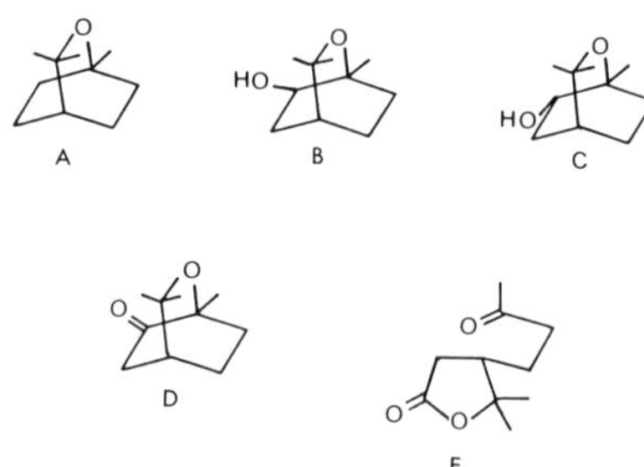

Fig. 5. Metabolites accumulated by *Pseudomonas flava* during growth on 1,8-cineole. Compounds are: (*A*) 1,8-cineole (growth substrate); (*B*) 2-*endo*-hydroxycineole; (*C*) 2-*exo*-hydroxycineole; (*D*) 2-oxocineole; (*E*) (R)-5,5-dimethyl-4-(3′-oxobutyl)-4,5-dihydrofuran-2(3H)-one. Data from MacRae, Alberts, Carman & Shaw (1979).

Metabolism of 1,8-cineole by Rhodococcus C1

Metabolite accumulation

Rhodococcus strain C1 was isolated by enrichment culture on 1,8-cineole (Wiliams et al. 1989) and grew in shake flasks with a doubling time of about 8 h. Gas Chromatographic (GC) analysis of diethyl ether extracts of neutral culture media showed the short-term accumulation of two compounds during the 15–18 h period of a 24 h growth regime. No acidic metabolites were detected. Sufficient quantities of the compounds for unequivocal identification and for metabolic studies were obtained from a 40 litre culture. One of these compounds gave a mass fragmentation pattern almost identical with that reported for the two isomers of 2-hydroxycineole and an ^{1}H-NMR spectrum indetical with that reported for 2-*endo*-hydrocycineole* (MacRae et al. 1979). These authors have reported that 2-*endo*-hydroxycineole is laevorotatory, $[\alpha]_D$-26° (c = 0.2 in ethanol) the alcohol accumulated by *Rhodococcus* C1 was dextrorotatory $[\alpha]_D$ + 27.5° in the same solvent. These result are only compatible with the compound being the optical isomer 6-*endo*-hydroxycineole. The second metabolite reacted with acidic 2,4-dinitrophenylhydrazine (Friedemann and Haugen, 1943) to give an orange insoluble 2,4-dinitrophenylhydrazone and a mass spectrum almost identical with that reported for 2-oxocineole (MacRae et al. 1979). The ^{1}H-NMR spectrum displayed all the diagnostic features attributable to 2-oxocineole. However, while 2-oxocineole is dextrorotatory the ketone accumulated by *Rhodococcus* C1 was laevorotatory, compatible with the compound being 6-oxocineole.

Rhodococcus C1 was also capable of growth with 6-endo-hydroxycineole and 6-oxocineole as sole carbon sources and it is logical to conclude that, by attacking the compound at carbon 6, it degrades 1,8-cineole in a manner that is analogous but enantiomeric to that utilized by *P. flava*. Further investigations of the route of 1,8-cineole degradation were made with subcellular systems.

1,8-Cineole hydroxylation

Although it is logical to assume that initial attack upon the molecule is by a monooxygenase that inserts an oxygen atom at the 6-*endo* position to form 6-*endo*-hydroxycineole we failed to identify any such hydroxylating system in subcellular preparations from induced cells. The use of a variety of different cell disruption procedures, inclusion of potential stabilizing agents and a range of electron donors and acceptors (in case the oxygen atom was derived from water) were all unsuccessful.

6-endo-hydroxycineole dehydrogenase

In contrast, an induced NAD+ linked 6-*endo*-hydroxycineole dehydrogenase was readily detected (specific activity 0.62 μmol/min.mg protein, pH optimum 10.5) and GC analysis showed 6-oxocineole to be the only product. The reverse reaction (specific activity 1.1 μmol/min.mg protein, pH optimum 7.5) yielded only 6-*endo*-hydroxycineole.

* Strict application of chemical nomenclature does not allow use of *endo* and *exo* in the (2,2,2)bicyclo system in relation to 1,8-cineole derivatives. To simplify presentation of results we have (a) designated as *endo* those alcohols in which the hydroxyl substitution is on the opposite side of the reference plane to the lowest priority bridge (see Fig. 5) and (b) made use of the numbering system consistent with the trivial name 1,8-cineole.

Further metabolism of 6-oxocineole

Not infrequently the catabolic transformation of cyclic hydrocarbons and alcohols into cyclic ketones is a preparation for ring oxygen insertion by a biological Baeyer-Villiger reaction (see Trudgill 1984, 1986) for reviews. These enzymes are flavoproteins and display typical monooxygenase characteristics, requiring an electron donor and incorporating one atom of oxygen into the organic product. When crude extract of 1,8-cineole-grown *Rhodococcus* C1 was incubated with 6-oxocineole and NADPH an oxygen-dependent oxidation of the NADPH was observed. Polarographic assays in the presence of limited amounts of either NADPH or 6-oxocineole allowed an approximate (1 : 1 : 1) stoichiometry of the reaction to be established. GC analysis of diethyl ether extracts of acidified reaction mixtures showed that the 6-oxocineole had been converted into a single more polar metabolite. GC-mass spectral analysis of the product gave a spectrum identical with that reported for the lactone (R)-5,5-dimethyl-4-(3′-oxobutyl)-4,5-dihydrofuran-2(3H)-one isolated by MacRae et al. (1979) (Fig. 5). Although the initial products of biological Baeyer-Villiger oxygenation of cyclic ketones are lactones there is no way in which this particular lactone can be formed directly from 6-oxocineole and the identity of the initial product of oxygenation was therefore uncertain.

Reaction with partially purified 6-oxocineole oxygenase

Purification of the NADPH-linked 6-oxocineole oxygenase from *Rhodococcus* C1 was a valid approach to determining the identity of the immediate oxygenation product. Unfortunately the enzyme was not very stable and althought addition of ethanol (5% v/v) did have a stabilizing influence we were unable, using either conventional procedures or by taking advantage of the short processing times of fast protein liquid chromatography (FPLC), to purify the enzyme more than 5–7 fold without rapid loss of activity.

When 15 mg of partially purified enzyme were incubated with 60 μmol of 6-oxocineole and excess NADPH the consumption of 50 μmol of oxygen was observed. Direct ferric hydroxamate assay for lactones (Cain 1961) showed them to be absent. When the aqueous reaction mixture was acidified (pH $\simeq$ 1) and extracted with diethyl ether the ether phase contained 'ferric hydroxamate positive' material which co-chromatographed with the 5,5-dimethyl-4-(3′-oxobutyl)-4,5-dihydrofuran-2(3H)-one obtained with crude cell extracts. This reinforced the view that this lactone is an artifact of isolation and not the immediate product of oxygenation.

5,5-Dimethyl-4-(3′-oxobutyl)-4,5-dihydrofuran-2(3H)-one is the lactone of 3-(1-hydroxy-1-methylethyl)-6-heptanoic acid. 4-Hydroxy acids readily lactonize in acidic solution and organic solvents. Wallach (1895) made use of precisely this spontaneous ring closure of 3-(1-hydroxy-1-methylethyl)-6-heptanoic acid in his investigations into the structure of α-terpineol. The working hypothesis that this acid is a catabolic intermediate from which the lactone is formed during extraction procedures was supported by the observation that 3-(1-hydroxy-1-methyl-ethyl)-6-heptanoic acid, prepared by alkalie hydrolysis of an aqueous solution of the racemic lactone [synthesized chemically from α-terpineol essentially as described by MacRae et al. 1979)], would support growth of *Rhodococcus* C1 and was oxidized rapidly by 1,8-cineole-grown cells.

Chemical Baeyer-Villiger oxygenation of 6-oxocineole

Magnesium monoperoxyphthalate is an efficient chemical Baeyer-Villiger reagent that gives clean reaction products in good yield. When the 6-oxocineole, isolated from culture medium, was incubated with this reagent in dimethylformamide, followed by addition of water, acidification and extraction into diethyl ether, the reaction product had the same GC retention time as 5,5-dimethyl-4-(3′-oxobutyl)-4,5-dihydrofuran-2(3H)-one.

These observations clearly suggested that the initial step in 6-oxocineole degradation is indeed a biological Baeyer-Villiger oxygenation. Chemical Baeyer-Villiger reagents insert the ring oxygen in such a manner that, when there are two different alkyl groups either side of the carbonyl the more highly substituted carbon atom becomes bonded to the new oxygen. From this it is possible to predict

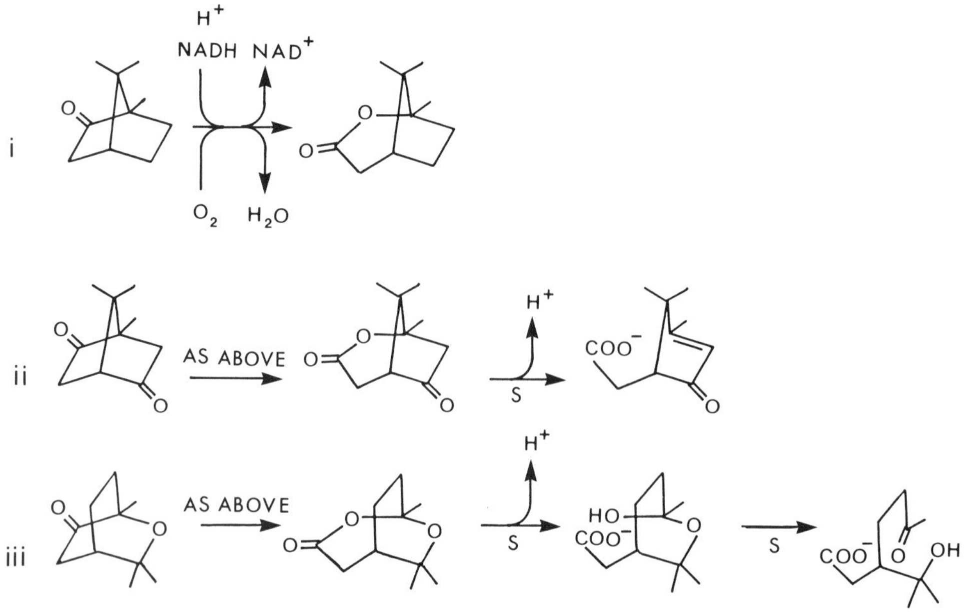

Fig. 6. Reactions catalysed by 2,5-diketocamphane 1,2-monooxygenase from *P. putida* ATCC 17453 with (*i*) (+)-camphor and (*ii*) 2,5-diketocamphane as substrates. The proposed oxygenation reaction with 6-oxocineole (*iii*) is followed by spontaneous cleavage to form the acyclic compound 3-(1-hydroxy-1-methylethyl)-6-oxoheptanoic acid. An alternative postulated spontaneous ring cleavage sequence leading to the same product is shown in Fig. 8. Adapted from Williams, Trudgill & Taylor (1989).

that the ring oxygenation product is 1,6,6-trimethyl-2,7-dioxabicyclo(3,2,2)-nonan-3-one. However, the steps between this lactone and 3-(1-hydroxy-1-methyethyl)-6-oxoheptanoic acid and any enzymes involved were not immediately clear.

Studies with 2,5-diketocamphane 1,2-monooxygenase and 6-oxocineole

2,5-Diketocamphane 1,2-monooxygenase of *P. putida* ATCC 17453 is a well described NADH-linked biological Baeyer-Villiger monooxygenase that is known to insert an oxygen atom into camphor between the carbonyl group and the bridgehead. The enzyme consists of NADH oxidase and oxygenating flavoprotein components that reversibly dissociate and which have been purified to homogeneity from (+)-camphor-grown cells (Taylor & Trudgill 1986). Although the enzyme is know to have a broad ketone substrate specificity it might be expected that 2-oxocineole would be preferred to 6-oxocineole. Preliminary experiments established that 6-oxocineole is a good substrate. This is presumably a result of the spatial symmetry of the (2,2,2) bicyclo skeleton (Fig. 6). Reactions in which 0.65 unit of enzyme was incubated with 10 μmol of 6-oxocineole and excess NADPH were used for product accumulation and analysis. Direct assay for lactones at the end of the reaction showed none to be present and, as occurred with the oxygenase preparations from *Rhodococcus* C1, acidification and diethyl ether extraction yielded, once again, the lactone 5,5-dimethyl-4-(3′oxobutyl)-4,5-dihydrofuran-2(3H)-one. This observation is of particular interest since camphor-grown *P. putida* ATCC 17453 is not known to have a lactone hydrolase. The clear implication is that the initial product of Baeyer-Villiger oxygenation is unstable. It was possible to confirm this fact in further investigations with a pH-stat. When the 2,5-diketocamphane 1,2-monooxygenase complex is incubated with (+)-camphor oxygenation takes place, a proton is consumed and the lactone product accumulates. In contrast the lactone formed from 2,5-diketocamphane is unstable and spontaneously ring opens to give an unsaturated acid so that overall proton balance is maintained. The experimental consequences of this are shown in Fig. 6. When 6-oxocineole is used as substrate proton balance is

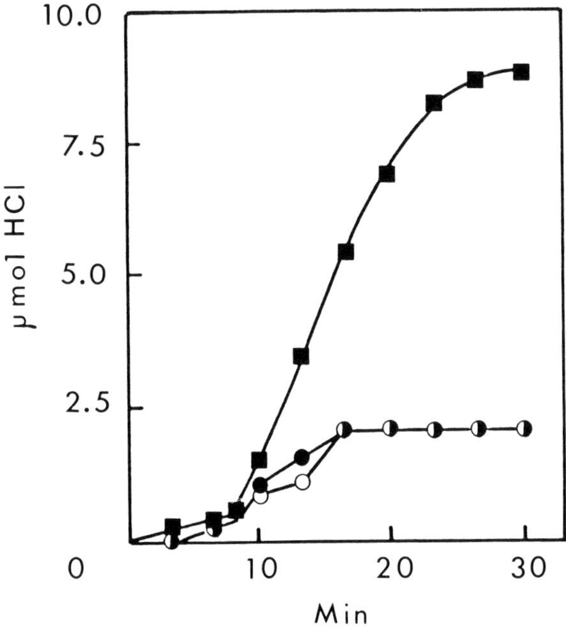

Fig. 7. A comparison of proton balance in the oxygenation reactions depicted in Fig. 6. The reaction vessel of a pH-stat contained 0.5 unit of 2,5-diketocamphane monooxygenase complex from *P. putida* ATCC 17453 and 10 μmol of NADH in 11 ml of distilled water adjusted to pH 6.7 with a minimum of phosphate buffer. After establishment of endogenous rates 10 μmol of (+)-camphor (■), 2,5-diketocamphane (●) or 6-oxocineole (○) were added and proton release monitored by controlled addition of 2 mM-HCl to maintain constant pH.
Adapted from Williams, Trudgill & Taylor (1989).

maintained confirming that, subsequent to the initial oxygenation step, further non-enzymic reaction(s) occur in which a proton is regenerated (Fig. 7).

Reaction sequence and ring cleavage

A reaction sequence for the cleavage of 6-oxocineole that is compatible with the experimental observations is shown in Fig. 8. The presumed lactone, 1,6,6-trimethyl-2,7-dioxabicyclo(3,2,2)nonan-3-one, is a strained molecule. We suggested originally that spontaneous cleavage is hydrolytic (Williams et al. 1989). This would form a hemiacetal with consequent further spontaneous cleavage of the ether linkage and formation of the anticipated acyclic acid. A possible alternative involving internal electron shifts would result in vinyl ether formation. Vinyl ethers are relatively unstable, especially in

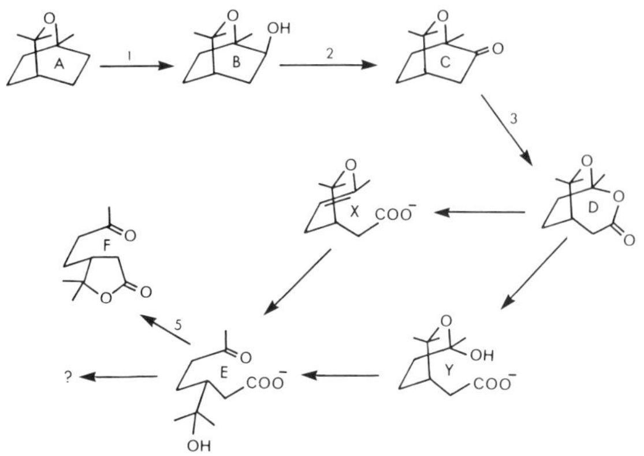

Fig. 8. Proposed ring cleavage reactions in the metabolism of 1,8-cineole by *Rhodococcus* strain C1. Compounds are: (*A*) 1,8-cineole; (*B*) 6-*endo*-hydroxycineole; (*C*) 6-oxocineole; (*D*) 1,6,6-trimethyl-2,7-dioxabicyclo(3,2,2)nonan-3-one; (*E*) 3-(1-hydrox-1-methylethyl)-6-oxoheptanoic acid; (*F*) 5,5-dimethyl-4-(3′-oxobutyl)-4,5-dihydrofuran-2(3H)-one. Compounds X, 2,6,6-trimethyl-5-acetyl-4,5-dihydropyran-2-ol and Y, 2,6,6-trimethyl-5-acetyltetrahydropyran-2-ol are proposed alternative transient intermediates of spontaneous ring cleavage. Enzymes are: (*1*) putative 1,8-cineole 6-*endo*-hydroxylase; (*2*) NAD-linked dehydrogenase; (*3*) NADPH-linked monooxygenase (a Baeyer-Villiger oxygenase). Reaction 5 is a spontaneous hydroxyacid ring closure (with loss of water) that occurs upon acidification of reactions and extraction with diethyl ether. Adapted from Williams, Trudgill & Taylor (1989).

dilute acidic solution, and spontaneous cleavage would again yield the anticipated acid. It is not known which of these alternative routes is followed.

The further metabolism of 3-(1-hydroxy-1-methylethyl)-6-oxoheptanoic acid has not been investigated but in this context it is of interest that 1,8-cineole-grown *Rhodococcus* C1 also oxidizes levulinic acid and acetone which can be derived theoretically from the open chain acid (Fig. 8) by fairly conventional metabolic steps.

MacRae et al. (1979) established the absolute configuration of the lactone (Fig. 5) accumulated in culture media by *P. flava*. Although we have not established the absolute configuration of the lactone formed by *Rhodococcus* C1 and 3-(1-hydroxy-1-methylethyl)-6-oxoheptanoic acid from which it is formed, logic would dictate that thay will have the (S) configuration.

Metabolism of 1,4-cineole

In principle the sequence of ring cleavage steps is also theoretically applicable to the less widely distributed isomer 1,4-cineole which, through a sequence of more highly strained intermediates, would yield 3-hydroxy-3-(1-methylethyl)-6-oxoheptanoic acid. In this context it is of interest that Rosazza et al. (1987) have reported 2-*endo*-hydroxy and 2-oxo derivatives as products of 1,4-cineole fermentation by a variety of bacteria and fungi grown in rich media.

Bacterial metabolism of α-pinene

Naturally occurring bicyclic monoterpenes are homocyclic and are fused 6/5 (borneol, camphor), 6/4 (α-pinene, β-pinene) or 6/3 (car-3-ene) carbon rings. The degradation of camphor has been studied in most detail and preparation for ring cleavage is dominated by hydroxylation and Baeyer-Villiger oxygenation reactions which form lactones although, in the case of *P. putida* ATCC 17453 lactone hydrolase activity has not been detected. (+)-Camphor-grown *P. putida* ATCC 17453 hydroxylates the compound at C5 and oxidises this to 2,5-diketocamphane. This appears to be a two-fold strategy; on the one hand it produces a molecule which is unstable and spontaneously ring opens once a biological Baeyer-Villiger oxygenase has inserted an oxygen between the keto group at C2 and the bridgehead, on the other it places a second keto group in such a position that the molecule can be manipulated, by formation of a CoA ester and a second ring oxygenation, so that a second spontaneous ring cleavage occurs (Fig. 3). Further metabolic studies on the degradation of the acyclic metabolite have not been reported.

Unlike camphor, in which the keto group might be expected to have a directing influence on catabolism, car-3-ene and α-pinene are unsaturated hydrocarbons. There is no significant published information on the microbial degradation of car-3-ene but more progress has been made in understanding the degradation of α-pinene.

Early studies of α-pinene metabolism by Pseudomonas strains

Shukla & Bhattacharyya (1968) and Shukla et al. (1968) isolated a variety of metabolites from culture medium when *Pseudomonas* strain PL was grown with α-pinene as sole carbon source. Accumulated acidic metabolites, primarily associated with oxidation of the C10 methyl group included perillic acid, which had previously been identified as a metabolite produced from limonene by this organism (Dhavaliker & Bhattacharyya 1966) and the acyclic acids 3-isopropylpimelic acid and 3-isopropenylpimelic acid. Although Shukla & Bhattacharyya (1968) presented, on the basis of accumulation experiments, a rather complex pattern of metabolic pathways leading from α-pinene to other compounds their evidence, including limited subcellular studies, is consistent with a primary degradative route that involves initial cleavage of the cyclobutane ring to form limonene or 1-*p*-menthene, oxidation of the C7 methyl group (C10 of α-pinene) to carboxyl and β-oxidative ring cleavage (Fig. 9).

In contrast to these observations Gibbon & Pirt (1971), Gibbon et al. (1972) and Tudroszen et al. (1977), again on the basis of metabolites accumulated by parent Pseudomonas strains and mutants, proposed a quite different catabolic route in which cleavage of the C6 ring occurred between carbon atoms 3 and 4 leading to the formation of 2-methyl-5-isopropylhexa-2,5-dienoic acid (Fig. 10). Although a biological Baeyer-Villiger oxygenation step was proposed no supporting subcellular evidence was reported.

α-Pinene metabolism by P. fluorescens NCIMB 11671

Recently Best et al. (1987) reported that extracts of α-pinene-grown *Pseudomonas fluorescens* NCIMB 11671 catalysed an NADH-linked consumption of oxygen in the presence of α-pinene. Rates of NADH and oxygen consumption suggested a 1 : 1 stoichiometry and the activity was located

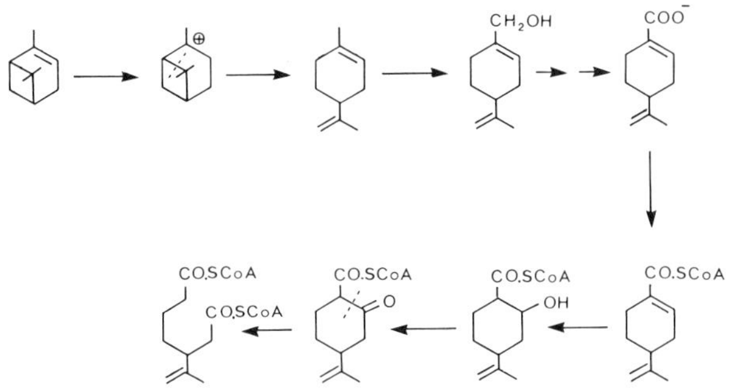

Fig. 9. Proposed pathway for cleavage of α-pinene by *Pseudomonas* PL. A prototrophic rearrangement of α-pinene to form limonene has been suggested, followed by methyl group oxidation and ring cleavage mediated by a β-oxidation cycle. Based on the results of Shukla & Bhattacharyya (1968) and Shukla, Moholay & Bhattacharyya (1968).

in the soluble protein fraction. The neutral reaction product was extracted into diethyl ether and mass spectral, ^1H and ^{13}C NMR analyses identified the compound as 2-methyl-5-isopropylhexa-2,5-dien-1-al, although the isomeric configuration was not reported. It is of particular interest that, on oxidation, this aldehyde would yield 2-methyl-5-isopropylhexa-2,5-dienoic acid, both isomers of which were reported (Gibbon & Pirt 1971; Gibbon et al. 1972; Tudroszen et al. 1977) as being accumulated from α-pinene by *Pseudomonas* PX1 (NCIMB 10684), *P. putida* PIN11 and mutants thereof.

A particularly remarkable feature was the rapid cleavage of both carbocyclic rings of α-pinene that was catalysed by very small amounts of protein. It was provisionally assumed that an initial hydroxylation of α-pinene occurred. However, the inclusion of atebrin in assays, as a supposed inhibitor of enzymes with flavin prosthetic groups and thus of a flavoprotein hydroxylase, resulted in an inhibition of ring cleavage and the accumulation of α-pinene epoxide. Subsequent experiments showed that

- the initial oxygenation step formed α-pinene epoxide,
- β-pinene and (+)-limonene were also substrates for the enzyme,
- the enzyme was sensitive to inhibitors that react with thiol groups and that complex Fe^{2+}.

Further preliminary studies with α-pinene epoxide

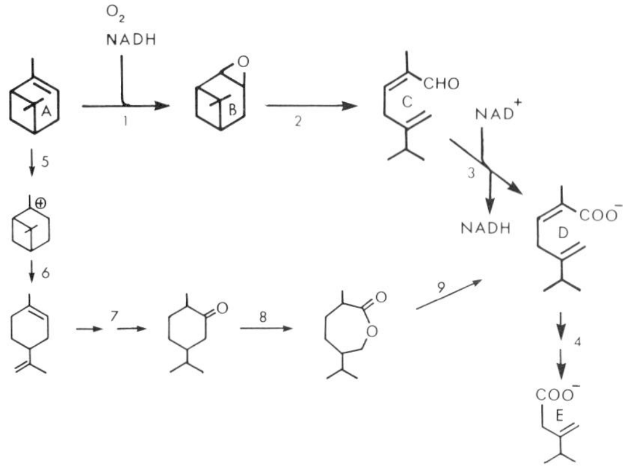

Fig. 10. Partial catabolism of α-pinene by *Nocardia* sp. strain P18.3, *P. fluorescens* NCIMB 11761 and *P. putida* NCIMB 10684. Compounds are: (*A*) α-pinene; (*B*) α-pinene epoxide; (*C*) *cis*-2-methyl-5-isopropylhexa-2,5-dienal; (*D*) *cis*-2-methyl-5-isopropylhexa-2,5-dienoic acid; (*E*) 3-isopropylbut-3-enoic acid. Reactions are: (*1*) NADH-dependent α-pinene monooxygenase (*P. putida* NCIMB 11671); (*2*) α-pinene epoxide lyase (all three organisms); (*3*) NAD-linked 2-methyl-5-isopropylhexa-2,5-dienal dehydrogenase (*Nocardia* P18.3); (*4*) proposed β-oxidation cycle to give the acid identified by Tudroszen, Kelly & Millis (1977). Reactions 5–9 were originally proposed for *P. putida* NCIMB 10684 by Gibbon & Pirt (1971).

showed that the soluble protein extract of α-pinene-grown *P. fluorescens* NCIMB 11671 catalysed a very rapid cleavage of α-pinene epoxide with the formation of the acyclic aldehyde 2-methyl-5-isopropylhexa-2,5-dienal.

α-Pinene metabolism by Nocardia sp. strain P18.3

We have isolated a Gram-positive organism *Nocardia* sp. strain P18.3 by elective culture with α-pinene. Cells grown on the terpene oxidize the growth substrate and α-pinene epoxide at comparable rates. Although we have been unable to demonstrate the putative epoxide-forming monooxygenase in this organism, incubation of extracts of α-pinene-grown cells with α-pinene epoxide resulted in its very rapid conversion into the *cis* isomer of 2-methyl-5-isopropylhexa-2,5-dienal [specific activity 16 μmol/min · mg protein] (Griffiths et al. 1987). Quite rapid transition to the *trans* isomer took place non-enzymically in the glycine NaOH

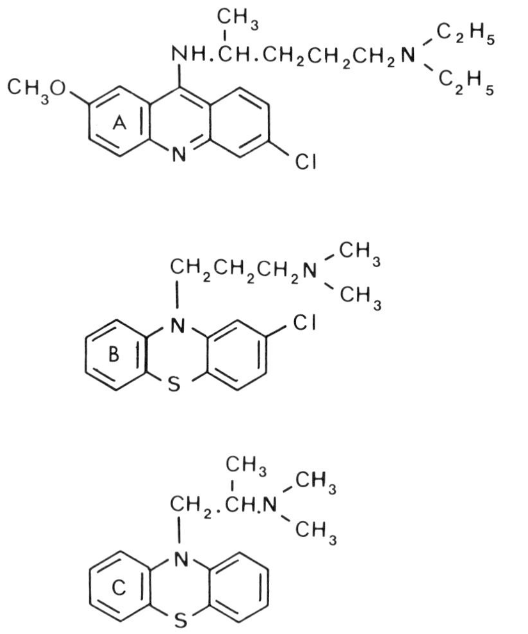

Fig. 12. Compounds that act as transition state inhibitors of α-pinene epoxide lyase. (*A*) atebrin (6-chloro-9-[(4-diethylamino)-1-methylbutyl]-amino-2-methoxyacridine); (*B*) chlorpromazine (2-chloro-10-(3-dimethylaminopropyl)-phenothiazine; (*C*) promethazine (10-(2-dimethylaminopropyl)-phenothiazine).

taking advantage of additional binding interactions of a transition state by incorporating key structural elements of it in the stable structure of the inhibitor. One result is that transition state analogue inhibitors may have a much higher affinity for the enzyme active site than traditional ground state inhibitors. This is exemplified, for example, by the inhibition of 2,3-oxidosqualene cyclases by 2-*aza*-2,3-dihydrosqualene, 2-*aza*-2,3-dihydrosqualene-*N*-oxide and several derivatives thereof that show obvious structural analogy with carbocation intermediates of the cyclization process (see Benveniste 1986 for review).

In the case of α-pinene epoxide lyase the precise structural analogy between the potent inhibitors and proposed carbocation intermediates formed in catalysis (Fig. 11) is not obvious. Structural variations exist but the most potent compounds have both a ring nitrogen and a dimethyl or diethyl substituted amine group. The situation is further complicated by the fact that the two ionizable nitrogen atoms of atebrin have pK_a values of 7.5 (aromatic) and 10.1 (diethylamine) but the compound is a less effective inhibitor at pH 7 than at pH 9. In conclusion, although we are uncertain as to exactly how these inhibitors mimic transition state carbocations were are encouraged in our proposed catalytic mechanism (Fig. 11) by the proven requirement for both a positive charge and a cyclic component if the inhibitor is to be optimally effective.

Acknowledgements

All workers who have contributed to our growing understanding of this complex metabolic field are gratefully acknowledged. It is their efforts that have made this review possible. The SERC and Unilever are gratefully thanked for funds provided to support our work on bicyclic monoterpene metabolism.

References

Benveniste P (1986) Sterol biosynthesis. In: Briggs WR, Jones RL & Walbot V (Eds) Ann. Rev. Plant Physiol. 37: 275–308, Annual Reviews Inc. Palo Alto, Calif.

Best DJ, Floyd NC, Magalhaes A, Burfield A & Rhodes PM (1987) Initial steps in the degradation of alpha-pinene by *Pseudomonas fluorescens* NCIMB 11671. Biocatalysis 1: 147–159

Bradshaw WH, Conrad HE, Corey EJ, Gunsalus IC & Lednicer D (1959) Microbial degradation of (+)-camphor. J. Amer. Chem. Soc. 81: 5507

Cain RB (1961) The metabolism of protocatechuic acid by a *Vibrio*. Biochem. J. 79: 298–312

Cantwell GG, Lau EP, Watts DS & Fall RR (1978) Biodegradation of acyclic isoprenoids by *Pseudomonas* species. J. Bacteriol. 135: 324–333

Carman RM, Macrae IC & Perkins MV (1986) The oxidation of 1,8-cineole by *Pseudomonas flava*. Aust. J. Chem. 39: 1739–1746

Chapman PJ, Meerman G, Gunsalus IC, Srinivasan R & Rinehart KL (1966) A new acyclic metabolite in camphor oxidation. J. Amer. Chem. Soc. 88: 618–619

Conrad HE, Dubus R & Gunsalus IC (1961) An enzyme system for cyclic ketone lactonization. Biochem. Biophys. Res. Commun. 6: 295–297

Conrad HE, Dubdus R, Namtvedt MJ & Gunsalus IC (1965) Mixed function oxidation. II. Separation and properties of the enzymes catalysing camphor lactonization. J. Biol. Chem. 240: 495–503

Conrad HE, Leib K & Gunsalus IC (1965) Mixed function oxidation. III. An electron transport complex in camphor lactonization. J. Biol. Chem. 240: 4029–4037

Dhavalikar RS & Bhattacharyya PK (1966) Microbial transformations of terpenes: Part VII-fermentation of limonene by a soil pseudomonad. Indian J. Biochem. 3: 144–157

Dhavalikar RS, Rangachari PN & Bhattacharyya PK (1966) Microbial transformations of terpenes: Part IX-pathways of limonene degradation by a soil pseudomonad. Indian J. Biochem. 3: 158–164

Friedemann T & Haugen GE (1943) Pyruvic acid II. The determination of ketoacids in blood and urine. J. Biol. Chem. 147: 415–422

Gibbon GH & Pirt SJ (1971) Degradation of *a*-pinene by *Pseudomonas* PX1. FEBS Lett. 18: 103–105

Gibbon GH, Millis NF & Pirt SJ (1972) Degradation of *a*-pinene by bacteria. In: Terui G (Ed) Fermentation Technology Today (pp 609–612). Proceedings of the 4th international fermentation symposium, Osaka, Japan

Griffiths ET, Bociek SM, Harries PC, Jeffcoat R, Sissons DJ & Trudgill PW (1987) Bacterial metabolism of *a*-pinene: Pathway from *a*-pinene oxide to acyclic metabolites in *Nocardia* sp. strain P18.3. J. Bacteriol. 169: 4972–4979

Griffiths ET, Harries PC, Jeffcoat R & Trudgill PW (1987) Purification and properties of *a*-pinene oxide lyase from *Nocardia* sp. strain P18.3. J. Bacteriol. 169: 4980–4983

Gunsalus IC & Lipscomb JD (1973) Structure and reactions of a microbal monooxygenase: The role of putidaredoxin. In: Lovenberg W (Ed) Iron-Sulfur Proteins, Vol 1 (pp 151–171). Academic Press, New York

Gunsalus IC & Marshall VP (1971) Monoterpene dissimilation: Chemical and genetic models. CRC Crit. Rev. Microbiol. 1: 291–310

Gunsalus IC, Meeks JR, Lipscomb JD, Debrunner P & Munck E (1974) Bacterial monooxygenases-the P450 cytochrome system. In: Hayaishi O (Ed) Molecular Mechanisms of Oxygen Activation (pp 559–613). Academic Press, New York

Macrae IC, Alberts V, Carman RM & Shaw IM (1979) Products of 1,8-cineole oxidation by a pseudomonad. Aust. J. Chem. 32: 917–922

Nishimura H, Noma Y & Mizutani J (1982) Eucalyptus as biomass. Novel compounds from microbial conversion of 1,8-cineole. Agric. Biol. Chem. 46: 2601–2604

Ougham HJ, Taylor DG & Trudgill PW (1983) Camphor revisited: Involvement of a unique monooxygenase in the metabolism of 2-oxo-Δ^3-4,5,5-trimethylcyclopentenylacetic acid by *Pseudomonas putida*. J. Bacteriol. 153: 140–152

Rosazza JPN, Steffens JJ, Sariaslani FS, Goswami A, Beale JM, Reeg S & Chapman R (1987) Microbial hydroxylation of 1,4-cineole. Appl. Environ. Microbiol. 53: 2482–2486

Seubert W (1960) Degradation of isoprenoid compounds by microorganisms. I. Isolation and characterization of an isoprenoid-degrading bacterium, *Pseudomonas citronellolis* n. sp. J. Bacteriol. 79: 426–434

Seubert W & Fass E (1964) Untersuchungen uber den bakteriellen Abbau von Isoprenoiden. V. Der Mechanismus des Isoprenoidabbaues. Biochem. Z. 341: 35–44

Seubert W & Remberger U (1963) Untersuchungen uber den bakteriellen Abbau von Isoprenoiden. II. Die Rolle der Kohlensaure. Biochem. Z. 338: 245–264

Seubert W, Fass E & Remberger U (1963) Untersuchungen uber den bakteriellen Abbau von Isoprenoiden. III. Reinigung und Eigenschaften der Gerenylcarboxylase. Biochem. Z. 338: 265–275

Shuka OP & Bhattacharyya PK (1968) Microbial transformation of terpenes: Part IX-pathways of degradation of *a*- and β-pinenes in a soil *pseudomonad* (PL-strain). Indian J. Biochem. 5: 92–101

Shukla OP, Moholay MN & Bhattacharyya PK (1968) Microbial transformations terpenes: Part X-Fermentation of *a*- and β-pinenes by a soil *pseudomonad* (PL-strain). Indian J. Biochem. 5: 79–91

Taylor DG & Trudgill PW (1986) Camphor revisited: Studies of 2,5-diketocamphane 1,2-monooxygenase from *Pseudomonas putida* ATCC 17453. J. Bacteriol. 165: 489–497

Trudgill PW (1984) Microbial degradation of the alicyclic ring. In: Gibson DT (Ed) Microbial Degradation of Organic Compounds (pp 131–180). Marcel Dekker, New York

Trudgill PW (1986) Terpenoid metabolism by *Pseudomonas*. In: Sokatch JR (Ed) The Bacteria, a Treatise on Structure and Function, Vol X (pp 483–525). Academic Press, New York

Trudgill PW, Dubus R & Gunsalus IC (1966a) Mixed function oxidation. V. Flavin interaction with a reduced diphosphopyridine nucleotide dehydrogenase, one of the enzymes participating in camphor lactonization. J. Biol. Chem. 241: 1194–1205

Trudgill PW, Dubus R & Gunsalus IC (1966b) Mixed function oxidation. VI. Purification of a tightly coupled electron transport complex in camphor lactonization. J. Biol. Chem. 241: 4288–4290

Tudroszen NJ, Kelly DP & Millis NF (1977) *a*-Pinene metabolism by *Pseudomonas putida*. Biochem. J. 168: 312–318

Wallach O (1895) Zur Constitutionsbestimmung des Terpineols. Chem. Ber. 28: 1755–1777

Williams DR, Trudgill PW & Taylor DG (1989) Metabolism of 1,8-cineole by a *Rhodococcus* species: Ring cleavage reactions. J. Gen. Microbiol. 135: 1957–1967

Yphantis DA (1964) Equilibrium ultracentrifugation of dilute solutions. Biochemistry 3: 297–317

Biodegradation **1**: 107–119, 1990.

Formation and physiological role of biosurfactants produced by hydrocarbon-utilizing microorganisms

Biosurfactants in hydrocarbon utilization

Rolf K. Hommel
Bereich Biochemie, Sektion Biowissenschaften, Karl-Marx-Universität, Talstr. 33, Leipzig, 0-7010, Germany

Key words: bacteria, bioemulsifiers, biosurfactants, glycolipids, n-Alkane degradation, yeasts

Abstract

Microbial growth on water-insoluble carbon sources such as hydrocarbons is accompanied by metabolic and structural alterations of the cell. The appearance of surface-active compounds (biosurfactants) in the culture medium or attached to the cell boundaries is often regarded as a prerequisite for initial interactions of hydrocarbons with the microbial cell. Under this point of view, biosurfactants produced by hydrocarbon-utilizing microorganisms, their structures and physico-chemical properties are reviewed. The production of such compounds is mostly connected with growth limitation in the late logarithmic and the stationary growth phase, in which specific enzymes are induced or derepressed. Addition of purified biosurfactants to microbial cultures resulted in inhibitory as well as in stimulatory effects on growth. Therefore, a more differentiated view of microbial production of surface-active compounds is proposed. Biosurfactants should not only be regarded as prerequisites of hydrocarbon uptake, but also as secondary metabolic products.

Introduction

One of the major problems to be overcome by microorganisms in hydrocarbon metabolization is to make the hydrophobic carbon source accessible to the cell. The high insolubility of alkanes in aqueous media necessitates specific uptake mechanisms. Mainly three different typs have been proposted:

- uptake of monodispered dissolved alkanes,
- direct contact of cells with large oil drops, and
- contact with fine oil droplets (pseudosolubilized alkane).

Whether mechanism is involved, alterations to the cell are required that will enable it to adhere to oil droplets and, in the latter case, to recognize the substrate to be solubilized. Einsele (1983) showed alterations of the cell surrounding of the yeast *Candida lipolytica* grown on n-hexadecane was accompanied by a stronger affinity to lipophilic carbon sources. Other examples were summarized by Boulton & Ratledge (1984).

Very small oil droplets (microemulsions) may be generated by mechanical means such as stirring, however those droplets will continously coalesce with each other. Stable submicron, pseudosolubilized alkane droplets may only be formed in presence of interfacial active substances. As reviewed by Haferburg et al. (1986) surfactants may promote growth of microorganisms on lipophilic substrates by improvement of hydrocarbon transport. Good growth was correlated in most cases with conditions which provide greater interfacial area. However, the results vary from strain to strain and depend on which surfactants are used. Hisatsuka et al. (1971) and Itoh et al. (1971, 1972) reported strong stimulatory effects of extracellular rhamnose lipid added in trace amounts to cultivations of the producing organism itself, *Pseudomonas aeruginosa*, growing on hydrocarbons. The ability of a rhamnolipid-negative mutant strain to grow on hydrocarbons was only restored after addition of rhamnoplipid (Itoh & Suzuki 1972). Based on this, the capacity of microorganisms able to utilize n-alkanes or other oils has been recognized mainly for production of interfacial active agents.

It has therefore been concluded that alkane-utilizing microorganisms produce such surfactant compounds to disperse the growth substrate into oil-in-water emulsions in order to increase the interfacial area and thereby enhance the availability of carbon source. In most recent reviews, the presence of biosurfactants has been regarded as indispensable for the microbial growth on water-immiscible substrates (Cooper & Zajic 1980; Boulton & Ratledge 1984; Lang & Wagner 1987; Syldatk et al. 1984; Syldatk & Wagner 1987, e.g.). From this point of view, this paper considers only biosurfactants, which are synthesized by bacteria and yeasts able to utilize hydrophobic carbon sources, together with some of their properties and conditions of product formation.

Types of biosurfactants

The presence of both a hydrophilic and a lipophilic moiety within the biosurfactant molecule creates the typical property of surfactant: depending on this concentration and conditions, surfactants may aggregate to form micelles or reversed micelles, and accumulate at liquid/liquid, liquid/gaseous and liquid/solid interfaces. The term 'biosurfactant' has been often used loosely to refer to other compounds like biopolymers which generally do not reduce interfacial tension but may prevent oil droplets from coalescing. The most representative bioemulsifier, Emulsan, recreted by *Acinetobacter calcoaceticus* growing on ethanol or alkanes has been the main subject of recent reviews of Gutnick & Shabtai (1987) and Gutnick & Minas (1987). Excretion of other polymeric bacterial bioemulsifiers were also reported from other bacteria growing not only on non-hydrocarbon substrates (Banerjee et al. 1983; Cooper & Goldenberg 1987, e.g.). The bioemulsifier from *C. lipolytica*, Liposan, appeared only in the stationary phase of growth on long-chain alkanes (Cirigliano & Carman 1985). Its composition, a mannoprotein, was similar to that isolated from *Saccaromyces cerevisiae*, and strains of other genera including *C. lipolytica* (Cameron et al. 1988). Bioemulsifiers in general displayed similar composition to the cell wall or capsules.

In contrast, some biosurfactants are of a definite structure (see Fig. 1). The lipophilic portion is usually the hydrocarbon (alkyl) tail of one or more fatty acids which may be saturated, unsaturated, hydroxylated or branched. This fatty acid is linked to the hydrophilic group by a glycosidic, ester or amide bond.

Based on the character of the hydrophilic moiety, biosurfactants may be grouped into the following classes: (1) glycolipids, (2) lipopeptides, (3) fatty acids, (4) phospholipids, (5) neutral lipids.

Most biosurfactants are either neutral or negatively charged. The anionic character is due to carboxylate groups. A small number of cationic biosurfactants contain amine functions. Table 1 (and see also Fig. 1) gives a survey on most important biosurfactants and their properties. Detailed structures of biosurfactants were extensively described by Haferburg et al. (1986), Lang & Wagner (1987) and Gutnick & Minas (1987). In most cultivations of some *Actinomycetes*, mixtures of different surface-active compounds were described which were mostly, but not completely characterized.

The most important group of biosurfactants produced by hydrocarbon-utilizing microorganisms are glycolipids. In the trehalose lipids of *Actinomycetes* and related bacteria, a-branched β-hydroxy fatty acids, known as corynomycolic acids, with 20 to 40 carbon atoms are esterified (mono- and diesters) with the sugar (e.g., Rapp et al. 1979; Kretschmer et al. 1982). Tetraesters of trehalose with succinate and fatty acids reflecting the carbon skeleton of the hydrocarbon substrate in 2,3,4,2′ positions are known from *Rhodococcus erytropolis* (Ristau & Wagner 1983). A pentasaccharide esterified with seven fatty acids and succinate was synthesized by *Nocardia corynebacteroides* SM 1 (Powalla et al. 1989). The carbohydrate backbone contains one α,α-trehalose unit which is substituted at the same position as the trehalose tetraester of *R. erytropolis*. This trehalose lipid was therefore, considered to be biosynthetic precursor of the oligosaccharide lipid. Trehalose moieties of mycolates may be replaced by other sugars (Göbbert et al. 1988; Itoh & Suzuki 1974; Li et al. 1984).

In rhamnolipids, rhamnose or a di-rhamnosyl unit (α-1,2 linked) is glycosidically linked with β-hydroxy decanoic acid, which itself is esterified with a second β-hydroxy decanoic acid. Syldatk et al. (1985a) isolated also two rhamnolipids with only one fatty acid. Rhamnolipids in which an additional decenoyl moiety was linked in 2′ and 2″ position with the mono- and di-rhamnosyl unit, respectively, were described by Yamaguchi et al. (1976).

In the sophorose lipids of *Candida* (*Torulopis*) yeasts, sophorose is glycosidically linked with a long chain (ω − 1)-hydroxy fatty acid which mostly reflects the backbone of the water-insoluble carbon source (Tulloch 1976). The free carboxylic group may be lactonized with the 4″ hydroxy group. Additionally, ω-hydroxy fatty acids were reported in crystalline lipids from *Torulopsis apicola* (Weber et al. 1990). The lipids are, in general, mixtures of different sophorose lipids which exhibit different degrees of acetylation in 6′ and 6″ position or possess unsaturated fatty acids. The main components account up to 60% (Asmer et al. 1988) or even more than 80% (Stüwer et al. 1987). Additional types of lactonization of 17-hydroxy octadecanoic acid, 1,6′ lactone and 1,6″ lactone, respectively were reported by Asmer et al. (1988). The extracellular sophorose lipid of *Candida bogoriensis* was characterized as 13-[(2′-*O*-β-D-glucopyranosyl-β-D-glucopyranosyl)oxy]docosanoic acid 6′, 6″-diacetate (Cutler & Light 1979).

In the mannosyl-erythritol lipids of *Candida* sp. B-7, the mannosyl moiety of the carbohydrate backbone was reported to be esterified manly in 2′ and 6′ position by a mixture of C_7 to C_{17} fatty acid (Kawashima et al. 1983). In 2′, 3′, 4′ and 6′ position the acetylation and acylation (C_8 to C_{14} fatty acids) was reported in the lipid from *Candida antarctica* (Kitamoto et al. 1990a).

Lipopeptides usually appear as mixtures of closely related compounds which show slight varia-

a

b

c

d

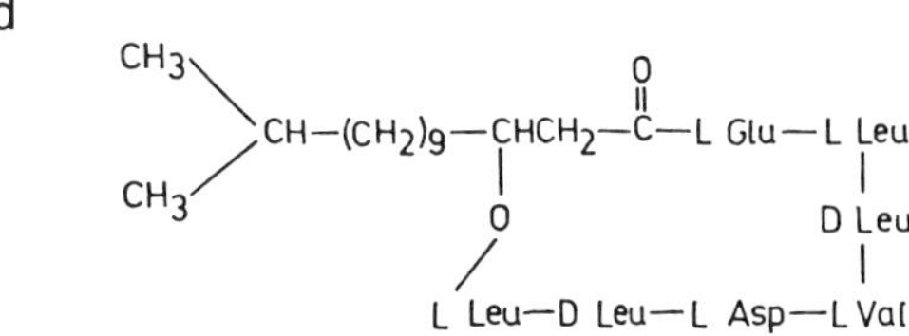

Fig. 1. Structures of selected biosurfactants.

(*a*) Rhamnolipids of *Pseudomonas aeruginosa.*

	I	II	III	IV
R_1	L-α-Rhamno-pyranosyl	H	L-α-Rhamno-pyranosyl	H
R_2	β-Hydroxy-decanoic acid	β-Hydroxy-decanoic acid	H	H

(*b*) Trehalose-6-monocorynomycolate of *Rhodococcus erytropolis*.

(*c*) Sophoroselipid (acid form) of *Torulopsis bombicola*.

(*d*) Surfactin of *Bacillus subtilis*.

tions in their amino acid composition and/or lipid portion which is mostly a hydroxy fatty acid. The lipopetides like surfactin from *Bacillus subtilis* (Vater 1986), viscosin of *Pseudomonas fluorescence* (Neu et al. 1990) or cyclodepsipeptides of *Serratia marcescens* (Matsuyama et al. 1986), represent a family of cyclic petides with consist of 8 to 17 amino acids.

Growth and biosurfactant formation

Mixtures of different biosurfactants

Mixtures of surface-active lipids have been seen to appear during growth of corynebacteria and related bacteria on hydrocarbons and/or kerosene (Gerson & Zajic 1979; MacDonald et al. 1981; Duvnjak & Kosarik 1985, e.g.). In some cases, the surface tension of broth was lowered before cell growth started (Cooper et al. 1979; MacDonald et al. 1981). Distinct maxima of surface activity of the culture broth during alkane cultivations were attributed to the sequentional appearance of free corynomycolic acids, lipopetides, phospholipids, neutral lipids and fatty alcohols. The corynomycolic acids were mainly constituents of some of more complex compounds such as lipopeptides (Cooper et al. 1979). Highest surfactant concentrations were measured in the medium of hexadecane growing on cultures (Atkit et al. 1981). Supplementation of glucose reduced biosurfactant yields. Duvnjak & Kosaric (1985) noted biosurfactant production by *Corynebacterium lepus* also on glucose, however, the surfactant remained cell-bound and could be released by treatment of cells with n-alkanes. It was concluded that the direct contact between cells and hexadecane was important for hydrocarbon transport. Similar observations have been made with *C. lepus* and with *Arthrobacter paraffineus* (Duvnjak et al. 1982). The excretion a peptidotrehaloselipid by the coryneform bacterium species H13A, was accompanied with the stationary phase of growth on alkanes (Singer 1985). On water-

Table 1. Selected properties of some structurally elucidated biosurfactants.

Type	Charge	Location	σ_s (mNm^{-1})	c.m.c.* (mg l^{-1})	σ_i (mNm^{-1})	Reference
Trehalose mycolates						Kretschmer et al. (1982)
mono-	nonionic	c.b.[1]	32	2	16	
di-	nonionic	c.b.	36	4	17	
Sugar esters of mycolates						Li et al. (1984)
Mannose	nonionic	c.b.	40	5	19	
Glucose	nonionic	c.b.	40	10	9	
Maltose	nonionic	c.b.	33	1	1	
Rhamnolipids						Syldatk et al. (1985a)
R I	anionic	c.f.[2]	31	20	<1	
R II	anionic	c.f.	25	200	<1	
R III	anionic	c.f.	31	20	3	
R IV	anionic	c.f.	30	200	<1	
Sophorolipids						
Lactone	nonionic	c.f.	35	60	9	Stüwer et al. (1987)
Mixture	nonionic	c.f.	25		<0.9	Hommel et al. (1987)
acidic	anionic	c.f.			3	Lang et al. (1984)
Surfactin	nonionic	c.f.	27	5	1	Cooper & Zajic (1980)

[1] Cell-bound
[2] Cell-free (extracellular)
* cmc = critical micelle concentration, σ_i = interfacial tension, σ_s = surface tension.

soluble carbon sources the glycolipid remained cell-associated.

Trehalose lipids and sugar mycolates

R. erythropolis growing on hydrocarbon displayed a biphasic growth pattern (Rapp et al. 1979). In the first phase, cells grew in the alkane droplets and cell-bound trehalose mycolates were produced at a constant rate – only 10% of the total trehalose dimycolates were released into the culture medium. After phase inversion at a precise ratio of cells/glycolipid/alkane, the hydrophobized cells aggregated in the aqueous phase which caused a limitation of oxygen transfer. Growth limitation coincided with the end of biosurfactant production.

Similar growth patterns were observed for the biosurfactant production of *Rhodococcus aurantiacus* (Ramsay et al. 1988). The production of glycolipids was accompanied with initial growth. After the exponential phase, linear growth, which was attributed to limitations in hydrocarbon transport, was observed without changes in the oxygen uptake rate or in biosurfactant production. Synthesis of trehalose tetraester by *R. erythropolis* started after depletion of nitrogen in the stationary growth phase (Ristau & Wagner 1983). Limitations of multivalent metal ions or the lowering of specific growth rate by temperature shifts may also initiate tetraester biosynthesis.

Low concentrations of other mycolates have also been recorded during fermentation. The production of the pentasaccharide ester by *N. corynebacteroides* SM 1 was 4-fold higher with cells growing on alkanes than with glucose (Powalla et al. 1989). Minor components synthesized were dicorynomycolate and monocorynomycolate. The production of glycolipids continued into the stationary phase after the nitrogen source was exhausted. The substitution of trehalose in trehalose mycolates by other sugars was obtained with strains of *Arthrobacter, Corynebacteria, Brevibacteria* and *Nocardia* isolated as hydrocarbon-utilizing bacteria (Itoh & Suzuki 1974; Suzuki et al. 1974) after growth on the respective sugar. The same replacement of the sugar moiety has also been described by Li et al. (1984) with resting cells of *Arthrobacter* sp. DSM 2567 which provided higher yields than in batch cultures. The corynomycolic acids remained unchanged. Its synthesis did not demand a supply of hydrocarbons. This reaction is catalysed by a highly regioselective enzyme of the particulate cell fraction which accepted mono-, oligo-saccarides, sugar alcohols and *p*-nitrophenyl derivatives as substrates (Göbbert et al. 1988).

Rhamnose lipids

Rhamnolipids of *P. aeruginosa* are formed from cultures growing on either hydrocarbons (Hisatsuka et al. 1971; Itoh et al. 1971; Syldatk et al. 1985a, b) or on water-soluble carbon sources (Haferburg et al. 1989; Guerra-Santos 1984, 1986; Reiling et al. 1986; Syldatk et al. 1985a, b). Limitations of one component of the growth medium, e.g. nitrogen, phosphorus or sulphur, were prerequisites for biosynthesis to commence. The C : N ratio of the medium constituents was mostly above 20. High concentrations of bivalent cations inhibited biosynthesis of the surfactant. Limitation of growth by iron caused a further increase in rhamnolipid production (Guerra-Santos et al. 1986; Haferburg et al. 1989; Syldatk et al. 1985a). A direct correlation of rhamnolipid production and specific activity of glutamine synthetase was established (Mulligan & Gibbs 1989). Ammonium and glutamine repressed both the non-growth associated rhamnolipid production and the glutamine synthase activity in the wild-type strain and also an overproducing chloramphenicol tolerant mutant (Mulligan et al. 1989). The enzyme was at maximum activity at the end of exponential phase, the time at which nitrogen and phosphate sources had been exhausted, and remained high throughout rhamnolipid production. Rhamnolipid production occurred in both batch cultivations and with resting cells (Syldatk et al. 1985a). Neither rhamnose nor β-hydroxy decanoic acid could be replaced indicating a *de novo* synthesis of this lipid.

Rhamnolipids may be produced in continuous culture (e.g. Guerra-Santos et al. 1984; Reiling et al. 1986) and also by alginate immobilized cells on glycerol (Müller-Hurtig et al. 1987) under specific limitations. In batch cultures, production was not growth-associated using either n-alkanes or water-soluble substrates (Syldatk et al. 1984). In contrast to this, Koronelli et al. (1983) reported the appearence of a peptidoglycolipid containing rhamnose by *P. aeruginosa* P-20 was strongly connected with the beginning and to the middle of exponential growth phase on hexadecane.

Sophorose lipids

Formation of extracellular sophorolipids by yeasts of the genus *Candida* and *Torulopsis* has been reviewed by Tulloch (1976) and Spencer et al. (1979). The latter genus has been recently reclassified. Producing strains like *Torulopsis magnoliae* (Gorin et al. 1961), *T. apicola* (Tulloch & Spencer 1968), *Torulopsis gropengiesseri* (Jones 1967) and *Torulopsis bombicloa* (Spencer et al. 1970) now belong to the genus *Candida*.

In presence of glucose, methylesters of saturated and unsaturated fatty acids in the range between C_{13} to C_{20} were directly incorporated into the surfactant molecule (Tulloch 1976). Fatty alcohols also served as substrates (Spencer et al. 1979). Ito et al. (1980) isolated a lactonic sophorose lipid after growth of *T. bombicola* KSM-36 on safflower oil and glucose. Under these conditions anionic glycolipids were also produced (Ito & Inoue 1982). With hydrocarbons as sole source of carbon, glycolipid production did not occur (Ito & Inoue 1982). Maximum production of an anionic glycolipid mixture of sophorose lipids was obtained with *T. bombicola* ATCC 22214 on a mixture of carbohydrate and vegetable oils or by subsequent addition of one of the principal carbon sources (Cooper & Paddock 1984). Under both conditions, most of the glycolipids appeared in the late exponential phase of growth. *T. bombicola* ATCC 22214 cultivated both on a mixture of glucose and oleic acid and sole on oleic acid produced a mixture of several different sophorosides of 17-hydroxy octadecanoic and 17-hydroxy octadecenoic acid (Asmer et al. 1988). In broths of logarithmicalty growing cells, only traces of glycolipid were detected. Glycolipid production started in the late exponential phase.

A mixture of water-soluble non-ionic glycolipids was obtained with *T. apicola* IMET 43734 (Hommel et al. 1987) with higher yields on hexadecane than on glucose. Biosurfactant concentration increased with increasing hexadecane concentrations. Excretion started in the middle of exponential growth phase and rose suddenly and very markedly in the late exponential phase. On a mixture of glucose and alkanes or vegetable oils, the same strain produced a crystalline glycolipid (Stüwer et al. 1987). This lipid was only obtained in fermentations in which organic acids (preferably citrate) were added which themselves did not permit growth. Glycolipid excretion started with the beginning stationary phase and continued at a constant rate. With resting cells of *T. bombicola* ATCC 22214 Göbbert et al. (1984) tried but failed to change the sugar moiety. 17-Hydroxy octadecenoic acid was the lipophilic constituent independently whether shorter-chain (C_{14}, C_{15}) alkanes were used as carbon sources.

The decreasing susceptibility of growth and glycolipid production of *T. apicola* towards cerulenin (an inhibitor of fatty acid biosynthesis) during growth on glucose and on mixed carbon source suggested that glucose is the growth substrate in mixed substrate cultivation in spite of synchronic disappearence of both glucose and hexadecane (Hommel et al. 1990). Whole cells of *Torulopsis* sp. strain 319–67 hydroxylated stereospecifically octadecenoic acid to the 17-L-hydroxy acid (Heinz et al. 1969). This reaction was carried out by a mixed-function oxidase of a particulate fraction of cells grown on a mixture of glucose and oil (Heinz et al. 1970). In *T. apicola* IMET 43747 cytochrome P-450 was detected after growth on glucose as well as on hexadecane and on a mixture of both appearing in the late exponential phase which coincided with the beginning of glycolipid production (Fig. 2) (Kleber et al. 1989). No repression of the cytochrome P-450 occurred in presence of glucose which is different to inducible monooxygenases involved in alkane hydroxylation

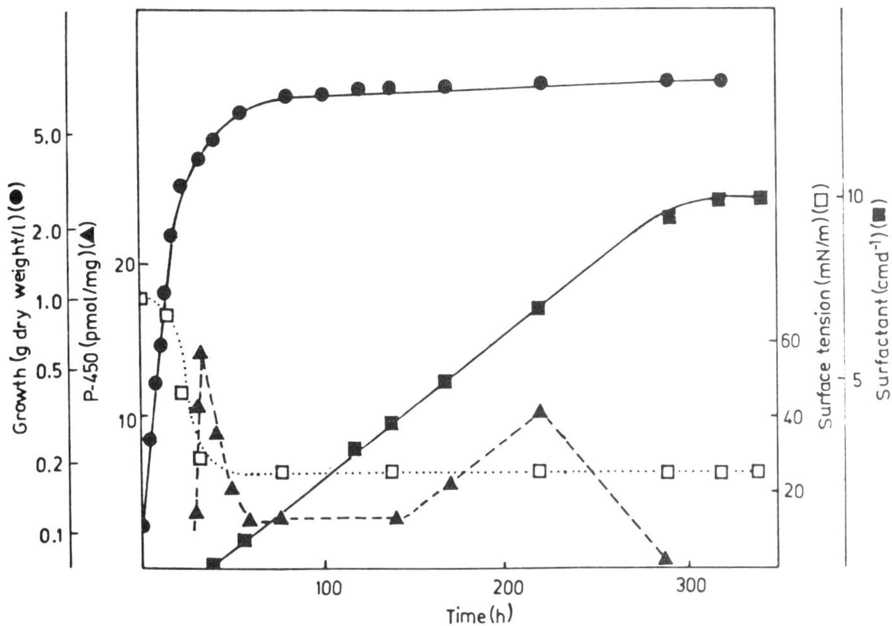

Fig. 2. Glycolipid production (critical micelle dilution, cmd^{-1}) and content of cytochrome P-450 during cultiviation of *Torulopsis apicola* IMET 43747 on glucose ($100 g l^{-1}$) (according to Kleber et al. 1989).

(Käppeli 1986). In the same context the fatty alcohol oxidase of *T. bombicola* ATCC 22214 was not repressed by glucose. The two constitutive enzymes of stationary cells (Hommel & Ratledge 1990) oxidized fatty alcohols and diols but ω-hydroxy fatty acids.

A sophoroside of a slightly different structure is formed by *C. bogoriensis* growing on glucose (Tulloch et al. 1968; Esders & Light 1972). The relative amount of glucose and yeast extract regulated the productivity: high concentrations of glucose promoted glycolipid biosynthesis (Cutler & Light 1979). The production was not growth associated and continued into the stationary phase of growth. Two different glycosyltransferases induced in the late exponential growth phase catalysed the stepwise transfer of UDP-glucose units to the hydroxy fatty acid and to 13-glucopyranosyloxydocosanoic acid, respectively (Breithaupt & Light 1982).

Mannosyl-erythritol lipids

The formation of an acylated mannosyl-erythritol lipid by *Candida* sp. B-7 is closely connected with the metabolism of n-alkanes and triacylglycerol (Kawashima et al. 1983). The addition of this lipid strongly increased the growth rate. *Candida* sp. KSM-1529, which is also able to produce this lipid, produced only non-acylated mannosyl-erithritol when grown on glucose and this probably serves as the direct precursor of the lipid (Kobayashi et al. 1987). Recently isolated strains of *Candida antarctica* were able to produce large amounts of a structurally modified mannosyl-erythritol lipid when grown on vegetable oils but not on alkanes (Kitamoto et al. 1990a). Supplementation of glucose did not affect the production. The addition of glycerol and erythritol caused significant increases in yields (Kitamoto et al. 1990b).

Lipopeptides

B. subtilis is only able to synthesize surfactin if grown on water-soluble carbon sources. The biosynthesis of peptide moiety of surfactin occurres non-ribosomally. Supplementation of hydrocarbons repressed surfactin synthesis whereas growth was not affected (Cooper 1984). The production of surfactin-like lipopeptides is not restricted just to *B. subtilis*. Similar componds have been described from other microorganisms which also belong to *Cyanobacteria* and *Actinomycetes* (Vater 1986). Production of surfactin started in the logarithmic growth phase and continued in the stationary phase (Cooper 1984; Vater 1986). Excessive production did not appear to be associated with actively growing cells; production was inversly correlated to biomass. Under both aerobic and anaerobic conditions, *Bacillus licheniformis* JF-2 produced a biosurfactant with properties similar to surfactin (Javaheri et al. 1985). The formation of further types of lipopeptides, such as the ornithine- and/or lysine-containing lipids and other types of biosurfactants, was reviewed by Haferburg et al. (1986) and Lang & Wagner (1987).

Biosurfactant additions in hydrocarbon utilization

The reports of Hisatsuka et al. (1971) and Itoh et al. (1971, 1972) demonstrated the stimulation of hydrocarbons metabolism by *P. aeruginosa* in presence of added rhamnolipid. *T. bombicola* ATCC 2217 and strain KSM-36 utilized a number of alkanes which were normally not assimilated in presence of lactonic sophorosides, methyl sophorose lipid or a mixture of sophorose lipids which had been produced during growth on glucose and safflower oil (Ito & Inoue 1982). Additionally, enhanced rates and yields of biomass were obtained. Synthetic surfactants could not replace the glycolipids. However, stimulation of growth by sophorose lipids was restricted to sophoroside-producing *Torulopsis* strains. Safflower lipid, lactonic sophoroside and also methyl sophoroside inhibited in general growth of other typical alkane-utilizing yeasts tested such as strains of *Candida* and *Pichia* on hexadexane (Ito & Inoue 1982; Ito et al. 1980). Similar results were obtained with the water-soluble, non-ionic glycolipid fraction of *T. apicola* IMET 43747 (Hommel et al. 1987). Its addition shorten the lag phase of particular alkane-utilizing bacteria and yeasts but displayed a decreasing effect on growth rate for strains tested.

Table 2 shows that the inhibitory effect of the lactonic sophoroside of *T. apicola* on different strains of *Candida* is restricted when they are using alkanes (Hommel et al. 1988). The low solubility of the lactonic sophoroside, which is also a main constituent of the safflower lipid, was judged to be responsible for the inhibitory effect because it exhibited no ability to emulsify hydrocarbons (Ito et al. 1980). Cooper & Paddock (1984) excluded this glycolipid as a promoter of alkane uptake because it did not stabilize alkane-in-water emulsions in

Table 2. Effect of sophoroselipid (1 g l^{-1}) on growth (A_{600}) of yeasts on glucose and on hexadecane (according to Hommel et al. 1988).

Strain	Carbon source			
	Glucose[1]		Hexadecane[2]	
	without Sophoroside	with Sophoroside	without Sophoroside	with Sophoroside
Candida				
C. boidinii H127	5.0	4.6	0.32	0.11
C. catenulata H173	4.1	3.7	1.8	0.15
C. maltosa H47	5.2	5.3	4.5	0.12
C. melinii H48	4.4	4.2	0.55	0.15
C. membranefaciens H93	4.6	4.6	0.60	0.09
C. mycoderma H259	3.5	3.8	3.8	0.16
C. paralopsilosis H260	2.8	2.6	2.5	2.4
C. pseudotropicalis H261	3.2	n.d.[3]	0.4	0.28
C. rugos H263	3.9	n.d.	0.52	0.32
C. sake H58	4.2	4.0	5.5	3.8
C. vini H96	4.6	4.0	4.9	2.9
Torulopsis				
T. apicola IMET 43747	6.0	5.9	4.8	4.6
T. bombicola ATCC 22214	5.2	5.5	4.1	4.8
T. candia Y127	8.2	7.2	5.5	0.2
T. famata H158	7.6	7.2	7.8	0.28
T. glabrata IMET 43548	7.4	6.8	5.9	0.58

[1] Glucose concentration: 50 g l^{-1}
[2] Hexadecane concentration: 20 g l^{-1}
[3] n.d. = not determined

vitro. However, Lang et al. (1984), Hommel et al. (1987) and Stüwer et al. (1987) reported strong interfacial activities with n-alkanes. This property is the basis of appearence of fine, dispersed non-coalesing alkane droplets in cultivation broths of *T. apicola* IMET 43747 and *T. bombicola* ATCC 22214.

Macroscopic emulsions may be considered as a possibility to facilitate uptake (and metabolism) of water-immiscible carbon sources. Each microorganism itself has developed individual mechanisms of environmental adaption. Microorganisms such as *C. lipolytica* or *A. calcoaceticus* which excrete bioemulsifiers were strongly inhibited by the addition of sophorosides (Ito & Inoue 1982; Hommel et al. 1987).

Syldatk et al. (1984) concluded that there were different mechanisms of emulsification with charged and neutral biosurfactants. Ionic biosurfactants, like the rhamnolipids or sophorolipids, emulsify (pseudosolubilize) the alkane which increases the surface area. Non-ionic biosurfactants (trehalose lipids) render the charged cell surface hydrophobic which should then facilitate the attachment and subsequent passive transport of alkanes into the cell (Ramsay et al. 1988; Rapp et al. 1979).

In this context, reports on effects of biosurfactants in oil polluted areas are valuable. Poremba et al. (1989) assumed the participation of cell-bound and microemulsion-forming extracellular biosurfactants in crude oil degradation by marine bacteria. Similar results have been reported for the biphasic degradation of a hydrocarbon mixture by a natural population of soil bacteria in a stirred reactor (Oberbremer & Müller-Hurtig 1989). Whereas in the first phase, water-soluble components were degraded, in the second phase the production of biosurfactants was observed which coincided with near complete exhaustion of alkanes. The glycolipids produced were trehalose tetraesters comparable to those reported by Ristau & Wagner (1983). Additionally, large amounts of a trehalose diester were produced which however remained cell-bound as described for *R. erythropolis* (Rapp et al. 1979). The addition of different biosurfactants to this model system caused a doubling of hydrocarbon degradation rate (Table 3) (Oberbremer et al. 1990). The adaption phases of the biphasic degradation were shorted and the extent of hydrocarbon degradation was enhanced. In the second phase, both the degradation of biosurfactants and the *de novo* synthesis of surface-active glycolipids was observed.

Conclusions

In contrast to the bulk number of alkane degrading microorganisms there is only a limited number of bacteria and yeasts known to produce interfacially active compounds. Most of them have been de-

Table 3. Effect of glycolipid addition on degradation efficiency in oil polluted soils (according to Oberbremer et al. 1990).

Cultivation	Hydrocarbon elemination		Degradation capacity [g hydrocarbon (kg soil)$^{-1}$d^{-1}]
	Duration [h]	Degree [%]	
With oxygen limitation			
Without	114	81	16.3
Trehalose-6,6′-dicorynomycolates	71	93	37.2
Sophorose lipids	75	97	39.0
Cellubiose lipids	79	99	32.3
Rhamnose lipids	77	94	28.2
Trehalose-2,3,4,2′-tetraesters	94	95	23.8
Without oxygen limitation			
Without	69	89	28.1
Sophorose lipids	57	95	46.5

scribed in connection with growth of microorganisms on hydrocarbons (Wagner et al. 1981; Zajic & Panchal 1976; Kosarik et al. 1983; Singer 1985). Effects of biosurfactants added to alkane cultivations have established that the function of these compounds in alkane emulsification or micelle formation is mainly restricted to the producing organism itself. The production of biosurfactants is, in general, not growth associated. However, already small amounts above critical micelle concentrations (c.m.c.) may allow the pseudosolubilization of such substrates. Indeed, the c.m.c. values reported (cf. Table 1 and Gutnick & Minas 1987) are generally very low. The substitution of trehalose in mycolic acid esters by other sugars suggests that alkanes do not induce their synthesis. In *Actinomycetes* their composition, partially or completely, corresponds to components of the cell wall or outer membrane. One effect of hydrocarbons on biosurfactant recovery is probably on extraction of lipids from the cell membranes (Cooper & Goldberg 1987; Duvnjak et al. 1982; Duvnjak & Kosaric 1985). Additionally, the same surface-active compounds are usually excreted by alkane-utilizing, biosurfactant producers like *P. aeruginosa* and strains of *Torulopsis* growing on water-soluble carbon sources. Biosurfactant production by *B. subtilis* is inhibited by alkanes (Cooper 1984) and is only connected with metabolism of water-soluble carbon sources. Similarly sophoroside production by *C. bogoriensis* is not linked to alkane utilization (Esders & Light 1972).

Common regulatory principles of biosurfactant synthesis are different and respond to different kinds of nutrient limitation (Syldatk & Wagner 1987). This indicates that enzymes involved in biosynthesis are normally repressed in actively, growing cells. The exhaustion, i.e. limitation, of either the carbon sources, or nitrogen or phosphorous is considered one factor leading to derepression of enzymes of secondary metabolism (Behal 1986; Malik 1982). The appearence and non-repression of cytochrome P-450 by glucose and the constitutive character of fatty alcohol oxidase in *T. apicola* and *T. bombicola* (Kleber et al. 1989; Hommel & Ratledge 1990), the time courses of glutamnine sythetase in *P. aeruginosa* (Mulligan & Gibbs 1989) and of the glycosyltransferases in *C. bogoriensis* (Breithaupt & Light 1982), suggests different regulatory mechanisms to those controlling primary metabolism.

In summary, a general consideration of biosurfactants as prerequiste of microbial alkane metabolism can not explain all the physiological aspects known so far. Some of biosurfactants, such as the sophorosides, rhamnolipids or peptidolipids, should be attributed to secondary metabolism, others, such as the trahalose esters or coynomycolic acids, are similar to existing cell wall constituents, and may be involved in cellular adaptation to hydrophobic growth substrates. Nevertheless, both types of surfactant may promote growth of natural mixed populations on hydrocarbon and some may also possess antimicrobial properties (Haferburg et al. 1986; Lang et al. 1989).

References

Asmer H-J, Lang S, Wagner F & Wray V (1988) Microbial production, structure elucidation and bioconversion of sophorose lipids. J Am Oil Chem Soc 65: 1460–1466

Atkit J, Cooper DG, Manninen KI & Zajic JE (1981) Investigation of potential biosurfactant production among phytopathogenic corynebacteria and related soil microbes. Current Microbiol 6: 145–150

Banerjee S, Duttagupta S & Chakrabartyam (1983) Production of emulsifying agent during growth of *Pseudomonas cepacia* with 2,4,5-trichlorophenoxyacetic acid. Arch Microbiol 135: 110–114

Behal V (1986) Enzymes of secondary metabolism in microorganisms. Trends Biochem Sci 11: 88–91

Boulton CA & Ratledge C (1984) The physiology of hydrocarbon-utilizing microorganisms. In: Wiseman A (Ed) Topics Enz Ferment Biotechnol, Vol 9 (pp 11–77). John Wiley & Sons, New York Chichester Brisbane Toronto

Breithaupt TB & Light RJ (1982) Affinity chromatography and further characterization of the glucosyltransferases involved in hydroxydocosanoic acid sophoroside production in *Candida bogoriensis*. J Biol Chem 257: 9622–9628

Cameron DR, Cooper DG & Neufeld RJ (1988) The mannoprotein of *Saccaromyces cerevisiae* is an effective bioemulsifier. Appl Environ Microbiol 54: 1420–1425

Cirigliano MC & Carman GM (1985) Purification and characterization of liposan, a bioemulsifier from *Candida lipolytica*. Appl Environ Microbiol 50: 846–850

Cooper DG (1984) Unusual aspects of biosurfactant production. In: Ratledge C, Dawson P & Rattray J (Eds) Biotechnology for the oils and fat industry. American Oil Chemists'

Society (Monogr. 11) (pp 281–287). American Oil Chemists Society, Champaign, Ill

Cooper DG & Goldenberg BG (1987) Surface-active agents from two *Bacillus* species. Appl Environ Microbiol 53: 224–229

Cooper DA & Paddock DA (1984) Production of a biosurfactant from *Torulopsis bombicola*. Appl Environ Microbiol 47: 173–176

Cooper DG, Zajic JE & Gerson DF (1979) Production of surface-active lipids by *Corynebacterium lepus*. Appl Environ Microbiol 37: 4–10

Cooper DA & Zajic JE (1980) Surface active compounds from microorganisms. Adv Appl Microbiol 26: 229–253

Cutler AJ & Light RJ (1979) Regulation of hydroxydocosanoic acid sophoroside production in *Candida bogoriensis* by the levels of glucose and yeast extract in the growth medium. J Biol Chem 254: 1994–1950

Duvnjak Z & Kosaric N (1985) Production and release of surfactant by *Corynebacterium lepus* in hydrocarbon and glucose media. Biotechnol Letters 7: 793–796

Duvnjak Z, Cooper DG & Kosaric N (1982) Production of surfactant by *Arthrobacter paraffineus* ATCC 19558. Biotechnol Bioeng 24: 165–175

Einsele A (1983) Biomass from higher n-alkanes. In: Rehm H-J & Reed G (Eds) Biotechnology, Vol 3 (pp 43–81). Verlag Chemie, Weinheim

Esders TW & Light RJ (1972) Characterization and in vivo production of three glycolipids from *Candida borgoriensis*. 13-Glucopyranosylglucopyranosyloxydocosanoic acid and its mono- and diacetylated derivatives. J Lipid Res 13: 663–671

Göbbert U, Lang S & Wagner S (1984) Sophorose lipid formation by resting cells of *Torulopsis bombicola*. Biotechnol Letters 6: 225–230

Göbbert U, Schmeichek A, Lang S & Wagner F (1988) Microbial transesterification of sugar-corynomycolates. J Am Oil Chem Soc 65: 1519–1525

Gorin PAJ, Spencer JFT & Tulloch AP (1961) Hydroxy fatty acid glycosides of sophorose from *Torulopis magnoliae*. Can J Chem 39: 846–895

Guerra-Santos L, Käppeli O & Fiechter A (1984) *Pseudomonas aeruginosa* biosurfactant production in continous culture with glucose as carbon source. Appl Environ Microbiol 48: 301–305

Guerra-Santos LH, Käppeli & & Fiechter A (1986) Dependence of *Pseudomonas aeruginosa* continous culture biosurfactant production on nutritional and environmental factors. Appl Microbiol Biotechnol 24: 443–448

Gutnick DL & Minas W (1987) Perspectives on microbial surfactants. Biochem Soc Trans 15: 22S–35S

Gutnick DL & Shabtai Y (1987) Exopolysaccharide bioemulsifier. In: Kosaric N, Cairns WL & Gray NCC (Eds) Surfactants Sciences Series. Biosurfactants and Biotechnology, Vol 25 (pp 211–246). Marcel Dekker, New York Basal

Haferburg D, Hommel R, Claus R & Kleber H-P (1986) Extracellular microbial lipids as biosurfactants. Adv Biochem Engin./Biotechnol 33: 53–93

Haferburg D, Hommel R & Kleber H-P (1989) Biotechnologie extrazellulärer microbieller Glycolipide. Wiss. Z. Univ. Leipzig, Math.-nat. wiss. Reihe 38: 303–311

Heinz E, Tulloch AP & Spencer JFT (1969) Stereospecific hydroxylation of long chain compounds by a species of *Torulopsis*. J Biol Chem 244: 882–888

Heinz E, Tulloch AP & Spencer JFT (1970) Hydroxylation of oleic acid by cell-free extracts of a species of *Torulopsis*. Biochim Biophys Acta 202: 49–55

Hisatsuka K, Nakahara T, Sano N & Yamada K (1971) Formation of rhamnolipid by *Pseudomonas aeruginosa* and its function in hydrocarbon fermentation. Agric Biol Chem 35: 686–692

Hommel R & Ratledge C (1990) Evidence for two fatty alcohol oxidases in the biosurfactant-producing yeast *Candida* (*Torulopsis*) *bombicola*. FEMS Microbiol Letters 70: 183–186

Hommel R, Stegner S, Ziebolz C, Weber L & Kleber H-P (1990) Effect of cerulenin on growth and glycolipid production of *Candida apicola*. Microbios Letters (in press)

Hommel R, Stüwer O, Stuber W, Haferburg D & Kleber H-P (1987) Production of water-soluble surface-active exolipids by *Torulopsis apicola*. Appl Microbiol Biotechnol 26: 199–205

Hommel R, Stüwer O, Weber L & Kleber H-P (1988) Structure and properties of glycolipids produced by *Torulopsis apicola*. 14th Int Congr Biochem Abstracts, Vol 1 (p 242)

Ito S & Inoue S (1982) Sophorolipids from *Torulopsis bombicola*: Possible relation to alkane uptake. Appl Environ Microbiol 43: 1278–1283

Ito S, Kinta M & Inoue S (1980) Growth of yeasts on n-alkanes: Inhibition by a lactonic sophorolipid produced by *Torulopsis bombicola*. Agric Biol Chem 44: 2221–2223

Itoh S, Honda H, Tomita F & Suzuki T (1971) Rhamnolipid produced by *Pseudomonas aeruginosa* grown on n-paraffin. J Antibiot 24: 855–859

Itoh S & Suzuki T (1972) Effects of rhamnolipids on growth of *Pseudomonas aeruginosa* mutant deficient in n-paraffin-utilizing ability. Agric Biol Chem 36: 2233–2235

Itoh S & Suzuki T (1974) Fructose-lipids of *Arthrobacter, Corynebacteria, Nocardia* and *Mycobacteria* grown on fructose. Agric Biol Chem 38: 1443–1449

Javaheri M, Jenneman E, McInerney MJ & Knapp RM (1985) Anaerobic production of a biosurfactant by *Bacillus licheniformis* JF-2. Appl Environ Microbiol 50: 698–700

Jones DG (1967) Novel macroscopic glycolipids from *Torulopsis gropengiesseri*. J Chem Soc C: 479–484

Käppeli O (1986) Cytochromes P-450 of yeasts. Microbiol Rev 50: 244–258

Kawashima H, Nakahara T, Oogaki M & Tabuchi T (1983) Extracellular production of a mannosyl-erythritol lipid of a mutant of *Candida* sp. from n-alkanes and triacylglycerols. J Ferment Technol 61: 143–148

Kitamoto D, Akiba S, Hioki C & Tabuchi T (1990a) Extracellular accumulation of mannosylerythritol lipids by a strain of *Candida antarctica*. Agric Biol Chem 54: 31–36

Kitamoto D, Haneishi K, Nakahara T & Tabuchi T (1990b)

Production of mannosylerythritol lipids by *Candida antarctica* from vegetable oils. Agric Biol Chem 54: 37–40

Kleber H-P, Asperger O, Stüwer O, Stüwer B & Hommel R (1989) Occurrence and regulation of cytochrome P-450 in *Torulopsis apicola*. In: Schuster I (Ed) Cytochrome P-450: Biochemistry and Biophysics (pp 169–172). Taylor & Francis, London, New York, Philadelphia

Kobayashi T, Ito S & Okamoto K (1987) Production of mannosylerythritol by *Candida* sp. KSM-1529. Agric Biol Chem 51: 1715–1716

Koronelli TV, Komarova TI & Denisov YV (1983) The chemical composition of *Pseudomonas aeruginosa* peptidoglycolipid and its role in the process of hydrocarbon assimilation. Microbioloiya 52: 767–770

Kosaric N, Gray NCC & Cairns WL (1983) Microbial emulsifiers and de-emulsifiers. In: Rehm H-J & Reed G (Eds) Biotechnology, Vol 3 (pp 575–592). Verlag Chemie, Weinheim

Kretschmer A, Bock H & Wagner F (1982) Chemical and physical characterization of interfacial-active lipids from *Rhodococcus erythropolis* grown on n-alkanes. Appl Environ Microbiol 44: 864–870

Lang S, Gilbon A, Syldatk C & Wagner F (1984) Comparison of interfacial active properties of glycolipis from microorganisms. In: Mittal KL & Lindman B (Eds) Surfactants in Solution, Vol 2 (pp 1365–1376). Plenum Press, New York

Lang S, Katsiwela E & Wagner F (1989) Antimicrobial effects of biosurfactants. Fat Sci Technol 9: 363–366

Lang S & Wagner F (1987) Structure and properties of biosurfactants. In: Kosaric N, Cairns WL & Gray WL (Eds) Surfactant Science Series. Biosurfactants and Biotechnology, Vol 25 (pp 21–45). Marcel Dekker, New York Basel

Li Z-Y, Lang S, Wagner F, Witte L & Wray V (1984) Formation and identification of interfacial-active glycolipids from resting microbial cells. Appl Environ Microbiol 48: 610–617

MacDonald CR, Cooper DG & Zajic JE (1981) Surface-active lipids from *Nocardia erythropolis* grown on hydrocarbons. Appl Environ Microbiol 41: 117–123

Malik VS (1982) Genetics and biotechnology of secondary metabolism. Adv Appl Microbiol 28: 27–115

Matsuyama T, Murakami T, Fujita M, Fujita S & Yano I (1986) Extracellular vesicle formation and biosurfactant production by *Serratia marcescens*. J Gen Microbiol 132: 865–875

Müller-Hurtig R, Matulovic U, Feige I & Wagner F (1987) Comparison of the formation of rhamnolipis with free and immobilized cells of *Pseudomonas* spec. DSM 2874 with glycerol as C-substrate. Proc. 4th European Congress on Biotechnology, Vol 2 (pp 257–260). Elsevier Sci Publ, Amsterdam

Mulligan CN & Gibbs BF (1989) Correlation of nitrogen metabolism with biosurfactant production by *Pseudomonas aeruginosa*. Appl Environ Microbiol 55: 3016–3019

Mulligan CN, Mahmourides G & Gibbs BF (1989) Biosurfactant production by a chloramphenicol-tolerant strain of *Pseudomonas aeruginosa*. J Biotechnol 12: 37–44

Neu TR, Härtner T & Poralla K (1990) Surface active properties of viscosin: a peptidolipid antibiotic. Appl Microbiol Biotechnol 32: 5188–5200

Oberbremer A & Müller-Hurtig R (1989) Aerobic stepwise hydrocarbon degradation and formation of biosurfactants by an original soil population in a stirred reactor. Appl Microbiol Biotechnol 31: 582–586

Oberbremer A, Müller-Hurtig R & Wagner F (1990) Effect of the addition of microbial surfactants on hydrocarbon degradation in a soil population in a stirred reactor. Appl Microbiol Biotechnol 32: 485–489

Powalla M, Lang S & Wray V (1989) Penta- and disaccharide lipid formation by *Nocardia corynebacteroides* grown on n-alkanes. Appl Microbiol Biotechnol 31: 473–479

Poremba K, Gunkel W, Lang S & Wagner F (1989) Mikrobieller Ölabbau im Meer. Biologie in unserer Zeit 19: 145–148

Ramsay B, McCarthy J, Guerra-Santos L, Käppeli O & Fiechter A (1988) Biosurfactant production and diauxic growth of *Rhodococcus aurantiacus* when using n-alkanes as the carbon source. Can. J. Microbiol 34: 1209–1212

Rapp P, Bock H, Wray V & Wagner F (1979) Formation, isolation and characterization of trehalose dimycolates from *Rhodococcus erythropolis* grown on n-alkanes. J Gen Microbiol 115: 491–503

Reiling HE, Thanei-Wyss U, Guerra-Santos LH, Hirt R, Käppeli O & Fiechter A (1986) Pilot plant production of rhamnolipid biosurfactant by *Pseudomonas aeruginosa*. Appl Environ Microbiol 51: 985–989

Ristau E & Wagner F (1983) Formation of novel anionic trehalosetetraesters from *Rhodococcus erythropolis* under growth limiting conditions. Biotechnol. Letters 5: 95–100

Singer ME (1985) Microbial biosurfactants. Microbes oil recovery 1: 19–38

Spencer JFT, Gorin PAJ & Tulloch AP (1970) *Torulopsis bombicola* sp. n. Antonie van Leeuwenhoek 36: 129–133

Spencer JFT, Spencer DM & Tulloch AP (1979) Extracellular glycolipids of yeasts. In: Rose AH (Ed) Economic Microbiology. Secondary Products of Metabolism, Vol 3 (pp 522–540). Academic Press, London New York San Francisco

Stüwer O, Hommel R, Haferburg D & Kleber H-P (1987) Production of crystalline surface-active glycolipids by a strain of *Torulopsis apicola*. J Biotechnol 6: 259–269

Suzuki T, Tanaka H & Itoh S (1974) Sucrose lipids of *Arthrobacter, Corynebacteria* and *Nocardia* grown on sucrose. Agr Biol Chem 38: 557–563

Syldatk C, Lang S, Matulovic U & Wagner F (1985a) Production of four interfacial active rhamnolipids from n-alkanes or glycerol by resting cells of *Pseudomonas* species DSM 2874. Z Naturforschung 40c: 61–67

Syldatk C, Lang S, Wagner F, Wray V & Witte L (1985b) Chemical and physical characterization of four interfacial-active rhamnolipis from *Pseudomonas* spec. DSM 2874 grown on n-alkanes. Z Naturforschung 40c: 51–60

Syldatk C, Matulovic U & Wagner F (1984) Biotenside – Neue Verfahren zur mikobiellen Herstellung grenzflächenaktiver, anionischer Glycolipide. Biotech Forum 1: 58–66

Syldatk C & Wagner F (1987) Production of biosurfactants. In:

Kosaric N, Cairns WL & Gray WL (Eds) Surfactant Science Series. Biosurfactants and Biotechnology, Vol 25 (pp 21–45). Marcel Dekker, New York Basel

Tulloch AP (1976) Structures of extracellular glycolipids produced by yeasts. In: Wittling LA (Ed) Glycolipid Methodology (pp 329–344). American Oil Chemists Society, Champaign Ill

Tulloch AP & Spencer JFT (1968) Fermentation of long-chain compounds by *Torulopsis apicola*. IV. Products from esters and hydrocarbons with 14 and 15 carbon atoms and from methyl palmitoate. Can J Chem 46: 1523–1528

Tulloch AP, Spencer JFT & Deinema MA (1968) A new hydroxy fatty acid sophoroside from *Candida bogoriensis*. Can J Chem 46: 345–348

Vater J (1986) Lipopetides, an attractive class of microbial surfactants. Progr Colloid Polymer Sci 72: 12–18

Wagner F, Bock H & Kretschmer A (1981) Gewinnung von Tensiden mit n-Alkane-oxidierenden Mikroorganismen. In: Lafferty RM (Ed) Fermentation (pp 181–192). Springer, Wien

Weber L, Stach J, Haufe G, Hommel R & Kleber H-P (1990) Elucidation of the structure of an unusual cyclic glycolipid from *Torulopsis apicola*. Carbohydrate Res 206: 13–19

Yamaguchi M, Sato M & Yamada K (1976) Microbial production of sugar lipids. Chem Ind 17: 741–742

Zajic JE & Panchal CL (1976) Bio-emulsifiers. Crit Rev Microbiol 5: 39–66

Biodegradation **1**: 121–132, 1990.

Microbial degradation of chelating agents used in detergents with special reference to nitrilotriacetic acid (NTA)

Thomas Egli, Matthias Bally & Thomas Uetz
Swiss Federal Institute for Water Resources and Water Pollution Control,
Swiss Federal Institutes of Technology, CH-8600 Dübendorf, Switzerland

Key words: complexing agents, detergent builders, wastewater treatment, NTA, EDTA, phosphonates

Abstract

The extensive use of phosphate-based detergents and agricultural fertilizers is one of the main causes of the world-wide eutrophication of rivers and lakes. To ameliorate such problems partial or total substitution of phosphates in laundry detergents by synthetic, non-phosphorus containing complexing agents is practiced in several countries. The physiological, biochemical and ecological aspects of the microbial degradation of the complexing agents most frequently used, such as polyphosphates, aminopolycarboxylates (especially of nitrilotriacetic acid), and phosphonates are reviewed.

Abbreviations: AODC – Acridine orange direct counts, ATMP – Aminotrimethylphosphonate, DTPA – Diethylenetriaminepentaacetate, DTPMP – Diethylenetriaminepentamethylphosphonate, EDTA – Ethylenediaminetetraacetate, EDTMP – Ethylenediaminetetramethylphosphonate, ED3A – Ethylenediaminetriacetate, HEDP – Hydroxyethylidenediphosphonate, HEDTA – Hydroxyethylethylenediaminetriacetate, IDA – Iminodiacetate, IFT – Immunofluorescence test, MW – Molecular weight, NTA – Nitrilotriacetate, PA – Polyacrylate, PHC – Polyhydroxycarboxylate, PMS – Phenazine methosulfate, SDS-PAGE – Sodium dodecylsulfate polyacrylamide gel electrophoresis, SPP – Tetrasodiumpyrophosphate, STP – Pentasodiumtriphosphate

Introduction

The washing of textiles is a complex process in which dirt (e.g., fats, proteins or inorganic and organic salt incrustations) is either emulsified or dissolved in the aqueous phase without damaging the textile fibers. To achieve this, detergents consisting of surface-active substances (tensides), bleaching agents and various additives (enzymes, perfumes etc.) are used. Furthermore, to prevent deposition of scale on both textiles and washing machine parts and to support performance of tensides, metal sequestering and/or chelating agents (so-called 'builders') have to be added because, in alkaline washing-liquid, bivalent metal ions such as calcium and magnesium form water insoluble salts with carbonate and washing active tensides (Jakobi et al. 1983). Historically, deposition of scale on textile fibers was circumvented by removing bivalent ions from the washing suds by precipitation with excess soap, sodium carbonate or sodium silicate (Berth et al. 1983; Jakobi et al. 1983). The invention of mechanical washing machines required adoption of a different strategy from precipitation, whereby Ca^{2+} and Mg^{2+} ions were kept in solution in the form of water soluble complexes during the washing process.

The compounds which have been employed for

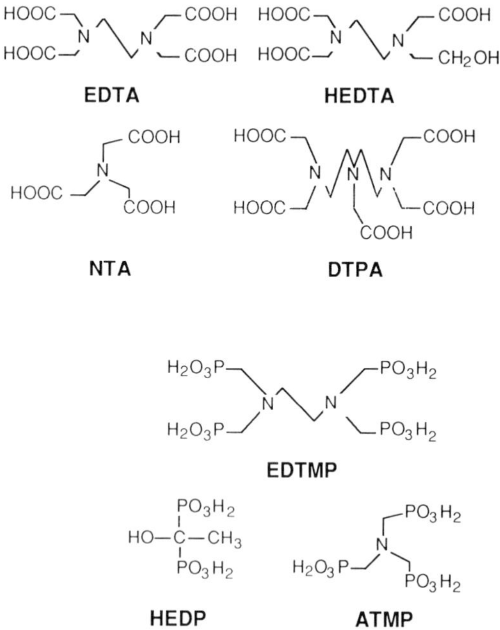

Fig. 1. Chemical structure of some synthetic chelating agents.

this purpose and which have been used in increasing amounts over the past 50 years were polyphosphates, mainly STP, because of its high complexing capacity, low toxicity and low price (Table 1). Over the last thirty years it has been shown in numerous studies that phosphates in domestic wastewater and agricultural fertilizers are the principal sources of phosphorus contributing to the widespread process of eutrophication of inland surface waters (Vollenweider 1968; Hauptausschuss Phosphate BRD 1978). Hence, in order to reduce the phosphorus load introduced into surface waters via wastewater various countries have implemented laws such as restriction of the phosphorus content in laundry detergents from typically 30–40% STP to a level of 4–5% STP (e.g. Canada, FRG, Italy) or the complete ban of phosphates from laundry detergents (e.g. the State of Indiana (USA) or Switzerland). This is because in wastewater usually a considerable part (40–70%) of the phosphorus stems from phosphate-based detergents, the remainder originating mainly from feces (Bunch & Ettinger 1967; Epstein 1972; Hauptausschuss Phosphate 1978).

Although worldwide the most frequently used chelating agent is still STP, alternative compounds have been suggested as potential builders (Kemper et al. 1975; Matzner et al. 1973) some of which are presently employed to partially or completely replace STP in detergents (Bernhardt 1990). The most important of them are the organic complexing agents NTA and citrate, but lately also silicates with cationic exchanging capacity such as Zeolites (Table 1). In addition to builders a range of other metal complexing agents such as phosphonates and polymeric carboxylic acids are frequently included in detergents in smaller quantities (1–2% in domestic laundry detergents) because of their ability to either slow down or inhibit crystal growth (Fig. 1).

Due to a number of facts (Mottola 1974; Tiedje 1980), the biodegradation of NTA has been the focus of considerable interest and, as a consequence, it is the synthetic chelating agent for which most information is available concerning both chemo-dynamic behaviour in the environment and biodegradation. This is probably due to the fact that it was essentially the first compound proposed as an efficient and cost-effective alternative to STP

Table 1. Estimated use of detergents and of chelating agents in the USA and Western Europe in 1981 (in 1000 metric tons). Data from Schneider (1984) and Egli (1988).

	USA	Western Europe
Detergents (total)[a]	7848[b]	7735
Synthetic laundry detergents	5209[b]	5643
Per capita consumption 1980	30.2[b,c]	18.9
Chelating agents		
STP	567	1100
SPP	19	11
EDTA	42	13.6
NTA	32	8.3
DTPA	4	0.5
HEDTA	18	2.0
Others	2.5	
Citric acid[d]	102	7.5
Gluconic acid	8	7.5
Organophosphonates	10	10

[a] Including soaps, soap based laundry detergents, synthetic laundry detergents and other cleaning agents.
[b] Figure given for North America.
[c] in kg
[d] Only approximately 30% used in detergents, rest is used in food and beverage industry.

in detergents. However, even today its environmental impact is assessed differently in different countries and whereas the use of NTA in laundry detergents is prohibited in Italy, it has been permitted in countries such as Canada, Sweden, Norway, Finland, Holland and Switzerland. Therefore, particular attention will be given in this review to various aspects of the biochemistry, physiology and ecology of the biodegradation of this aminopolycarboxylic acid. Although the carboxylic acids citrate and gluconate are frequently used as complexing agents in detergents for certain special purposes, they are common growth substrates and intermediates in microbial metabolism and they will not be considered here (for information see Egli 1988).

Nitrilotriacetic acid

NTA-degrading microorganisms

A wide range of microorganisms including *Pseudomonas, Bacillus* (Pickaver 1976), *Listeria* (Madson & Alexander 1985), and yeast species (Pickaver 1976) have been reported in the literature to be able to degrade NTA. Nevertheless, in the majority of detailed studies, Gram-negative motile, both obligately aerobic and facultatively denitrifying rods have been repeatedly isolated in pure culture primarily from soil and wastewater. Originally these isolates were assigned to the genus *Pseudomonas* (Cripps & Noble 1973; Tiedje et al. 1973; Focht & Joseph 1971; Kakii et al. 1986) and this led to the general view that NTA degradation in nature is primarily mediated by specialized *Pseudomonas* strains. However, in all cases this allocation was based on limited taxonomic information. A more detailed study in our laboratory in which a number of recently isolated obligately aerobic (Egli et al. 1988; Wehrli & Egli 1988) and one denitrifying strain (Wanner et al. 1990) were investigated, revealed that none of the isolates could be assigned to the genus *Pseudomonas* (as defined by de Vos & de Ley (1983)). Presently three distinctly different groups of Gram-negative, NTA-utilizing bacteria can be recognized and their key characteristics are given in Table 2. In addition to the information given in this table concerning the taxonomic characterization of these isolates, SDS-PAGE of soluble proteins, serological cross reactions, DNA: DNA-hybridization studies and rapid sequencing of 16S rRNA (El-Banna 1989; Wanner et al. 1990; El-Banna et al. unpubl.) all suggest that the three groups of Gram-negative, NTA-degrading bacteria are representatives of at least two, if not three, new genera. To date the phylogenetic position of the new genus consisting of strains 4–10 together with

Table 2. Key characteristics of the three different groups of NTA-degrading Gram-negative bacteria (data from Egli et al. 1988; Wanner et al. 1990, and El-Banna 1989).

Strains	Morphology	Sugars utilized	Respiration	G + C ratio	Main quinone	Main polyamine
TE 4–10, ATCC 27109, ATCC 29600	Motile rods, mostly pleomorphic	+	obligately aerobic	62.1–63.4	Q-10	*s*HSPD
TE 1, 2	Short rods or diplococci, non-motile, S-layer	–	obligately aerobic	63.3–63.5	Q-10	*s*HSPD
TE 11	Motile rods	+	facultatively denitrifying	65.3	Q-8	SPD

G + C ratio is given in mol% guanine plus cytosine; Q-10(9,8), Ubiquinone with 10(9,8) isoprene units in the side chain; *s*HSPD, *sym*-homospermidine; SPD, spermidine. For comparison: *Pseudomonas fluorescens* has a G + C ratio of 60.2 and contains mainly putrescine and Q-9 (Auling G & Busse J, pers. comm.).

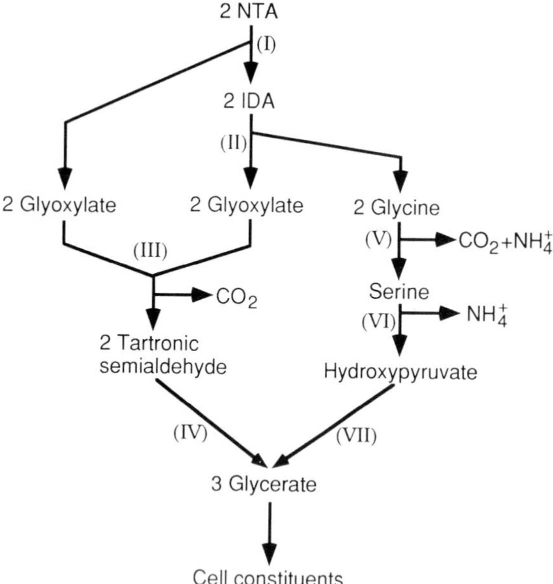

Fig. 2. Metabolic pathway for nitrilotriacetic acid proposed by Cripps & Noble (1973) for the Gram-negative obligate aerobic strain T23. I, Nitrilotriacetic acid mono-oxygenase; II, Unknown enzyme; III, Glyoxylate carboligase; IV, Tartronic semialdehyde reductase; V, Glycine decarboxylase and serine hydroxymethyltransferase; VI, Serine oxaloacetate aminotransferase; VII, Hydroxypyruvate reductase.

the two strains from the ATCC is best defined by a location in the α-2 branch of *Proteobacteria* where it is separated from its closest phytogenetic neighbours in the Rhizobia/Agrobacteria branch by a similarity coefficient (S_{AB}) of 0.59 (G. Auling, E. Stackebrandt, pers. comm.). There is also strong evidence that the second group consisting of strains TE 1 and 2 are representatives of a new genus in the α-branch of *Proteobacteria*. However, its exact taxonomic position in the α-2 branch still remains to be elucidated. Although, except for its ability to denitrify, no striking differences were observed between isolate TE 11 and the group of motile Gram-negative rods with respect to nutritional and physiological properties (Egli et al. 1988; Wanner et al. 1990), the presence of ubiquinone Q-8 and the polyamine spermidine clearly suggests a distant relationship to the group of obligately aerobic NTA-degraders and indicates that it should be allocated to the γ-subgroup of *Proteobacteria*. The combined presence of Q-8 and spermidine suggest allocation to the genus *Xanthomonas* (Ikemoto et al. 1980; Webb & G. Auling, pers. comm.). However, a number of physiological properties that are different to those typically exhibited by *Xanthomonas* spp. makes such an allocation of strain TE 11 rather unlikely and the possibility remains that a new genus will have to be established (Wanner et al. 1990).

Although none of the other strains isolated and described in the literature has been characterized to such an extent, the information available suggests that they may be members of the three groups of NTA-degraders described above. Furthermore, information presently available indicates that the ability to utilize NTA is not restricted to either a special genus or even a species but that it is a feature exhibited at least by members of both the α- and the γ-group of *Proteobacteria*. Further, a NTA-utilizing Gram-positive member of the *Rhodococcus* group has been isolated (Egli et al. 1988), thereby indicating that the ability to utilize NTA is not restricted to the Gram-negative bacteria.

Biochemistry

The biochemical pathway for NTA degradation was first investigated in two virtually physiologically and probably taxonomically identical bacteria, namely the obligately aerobic Gram-negative strains T23 (Cripps & Noble 1973) and ATCC 29600 (Firestone & Tiedje 1978). For both bacteria a mono-oxygenase was reported to be responsible for the conversion of NTA to IDA and glyoxylate. It was impossible to demonstrate further degradation of IDA in cell-free extracts. Based on enzyme induction patterns in NTA- and glucose-grown cells, Cripps & Noble (1973) proposed the biochemical pathway shown in Fig. 2. In an attempt to purify NTA mono-oxygenase, a protein, exhibiting NTA-stimulated NADH oxidation but completely devoid of NTA-splitting activity, was isolated from ATCC 29600 (Firestone et al. 1978; Snozzi & Egli 1987). These results indicate that NTA mono-oxygenase is a multi-component enzyme. This has been confirmed recently by the isolation of a functional enzyme consisting of two proteins, a flavin-containing component B with a MW of 32000 on

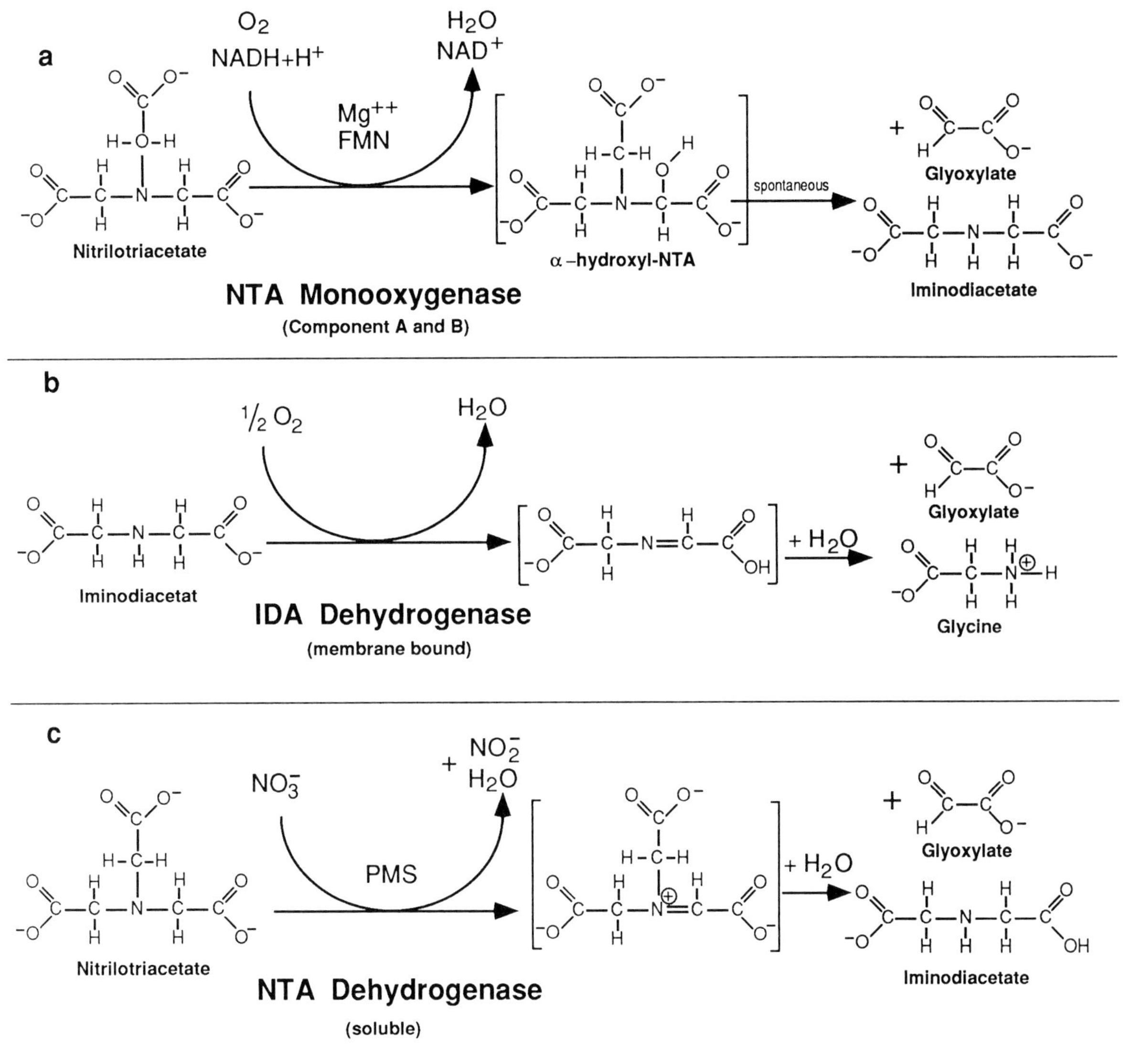

Fig. 3. Reactions catalyzed by nitrilotriacetate mono-oxygenase (*a*) and the membrane-bound iminodiacetate dehydrogenase (*b*) in the Gram-negative obligate aerobic strain ATCC 29600 and of nitrilotriacetate dehydrogenase (*c*) in the denitrifying strain TE 11.

SDS-PAGE exhibiting NTA/Mg^{2+}-stimulated NADH oxidation and a second component A (MW 50000 on SDS-PAGE) for which presently no catalytic activity is known. Catalytic activity, i.e., the formation of glyoxylate and IDA from NTA (Fig. 3a) was obtained only in the presence of components A and B, FMN and magnesium ions. Furthermore, the purified NTA mono-oxygenase did not accept IDA as a substrate (T. Uetz, unpubl.). This indicated that the subsequent metabolism of IDA was catalyzed by an additional enzyme and not by NTA mono-oxygenase as previously suggested by Firestone et al. (1978). Recently, IDA oxidation in the membrane fraction of NTA-grown cells of strain ATCC 29600 has been demonstrated (Fig. 3b). Glyoxylate was identified as a product of the reaction which was dependent on molecular oxygen and was inhibited by potassium cyanide (T. Uetz, unpubl. results). Whether NTA is degraded via NTA mono-oxygenase and IDA dehydrogenase in all obligately aerobic, Gram-negative isolates is not yet known.

It is obvious that in the facultatively denitrifying isolate TE 11 the metabolism of NTA must proceed via a different pathway during growth in the absence of molecular oxygen. In this bacterium, activity of a soluble NTA dehydrogenase which was dependent on the presence of PMS and nitrate was detected in membrane-free cell extract. In the reaction catalyzed (Fig. 3c) NTA was broken down stochiometrically to IDA and glyoxylate with the concommitant reduction of nitrate to nitrite (Wanner et al. 1989). Whether this dehydrogenase is also involved in the breakdown of NTA in aerobically grown cells and the fate of IDA in strain TE 11 still await elucidation.

To date, little biochemical information is available on the transport of NTA across the cell membrane. Wong et al. (1973) observed that transport of ^{14}C-labelled NTA was inhibited by both cyanide and azide which indicates that it is energy dependent. Experiments reported by Firestone & Tiedje (1975), in which the oxygen consumption of whole cells of strain ATCC 29600 incubated in the presence of various NTA-metal chelates was measured showed that this was highest when NTA was supplied in the presence of magnesium, manganese, calcium, iron, or sodium, whereas with nickel the rate of oxygen consumption was negligible. This suggests that transport of NTA into the cell is influenced by the nature of the metal-NTA complex. However, it is not yet known whether NTA is transported across the membrane in the form of a metal-NTA complex or whether before or during the process of transport the initially complexed metal ion has to be exchanged by other cations. In any case, the complexed cation being transported across the cytoplasmic membrane together with NTA has to be excreted to avoid intracellular accumulation.

Regulation of NTA catabolic enzymes

As has been shown in the previous section not all enzymes involved in the degradation of NTA have been identified. Nevertheless, the data presented by Cripps & Noble (1973) clearly demonstrated that growth on NTA led to enhanced activities of a series of enzymes in cells of strain TE23, including NTA mono-oxygenase and several enzymes involved in the transformation of glyoxylate and glycine to glycerate (compare Fig. 2). In our laboratory, the specific activity of NTA mono-oxygenase was recently measured as a function of both specific growth rate and of the nature of the growth-limiting substrate in chemostat cultures of strain ATCC 29600 (Wilberg 1989). It was found that the specific activity of NTA mono-oxygenase increased with decreasing specific growth rates during growth with NTA as the sole source of carbon, nitrogen and energy (Fig. 4). In addition, derepression of the synthesis of this enzyme was also observed at low growth rates when the cells were cultivated with glucose and ammonium. These results, shown in Fig. 4, suggest that although NTA is not absolutely necessary for expression, it stimulates additional induction of the synthesis of this enzyme. However, several questions with respect to the regulation of the synthesis of enzymes involved in NTA metabolism remain to be elucidated. Of particular interest are studies on the expression of the transport system for NTA, of NTA mono-oxygenase and of IDA dehydrogenase under different growth conditions, such as pulses of NTA to cultures exponentially growing with substrates other than NTA or the effect of different carbon/nitrogen substrates and ratios.

Ecological aspects of NTA degradation

The two crucial questions which remain to be answered with respect to NTA degradation under environmental conditions are firstly, how many and what kind of microorganisms capable of potentially degrading NTA are present in different ecosystems and secondly, how is the capacity to degrade NTA regulated in these microorganisms during growth in such systems.

Two different methods have been reported in the literature for the enumeration of NTA-degrading microorganisms in different ecosystems. Larson and co-workers used a ^{14}C-most-probable-number technique for enumeration of NTA-degrading microorganisms in estuarine water samples, a method which potentially allows enumeration of all organisms able to grow with NTA. In samples of different salinity the fraction of NTA-degrading microorganisms ranged from 0.0002 to 0.26% of the total bacterial cell count as measured by AODC (Larson & Ventullo 1986; Pfaender et al. 1985). Their results (Table 3) indicated that the distribution of NTA-degrading microorganisms was influenced neither by salinity nor by the dissolved organic carbon concentration in the water and they concluded that NTA degraders were indigenous members of the estuarine microbial community. In our own laboratory we used antibodies raised against whole cells of two different NTA-degrading isolates, namely strains ATCC 29600 and TE 2, in an indirect immunofluorescence assay to microscopically enumerate positively reacting cells in both wastewater treatment plants and natural waters (Wilberg 1989; Kemmler & Egli 1990). In this study it was found that the fraction of IFT-positive cells in surface waters was in the range of 0.01–0.1% of the total cell number assessed with AODC, whereas in activated sludge this number ranged from 0.1–1.0%, suggesting a ten-fold enrichment of NTA-degrading bacteria in treatment plants (Table 3). No significant difference in the fraction of IFT-positive cells was observed between samples collected from treatment plants which differed considerably with respect to their efficiency in NTA elimination. It should be stressed that these results might be influenced by both cross reactivity of the antisera used with non-NTA-degrading bacteria and that the sera might not recognize all NTA degraders present in the samples. This question could be answered by using a method in which sera of high specificity and uptake of ^{14}C-labelled NTA were combined.

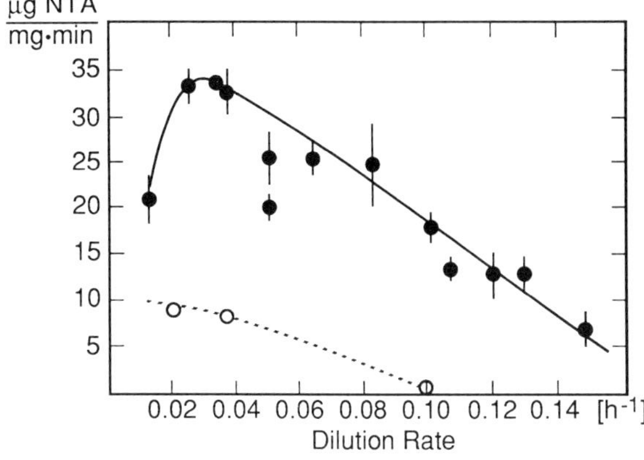

Fig. 4. Specific activity of nitrilotriacetate mono-oxygenase in cell-free extracts of strain ATCC 29600 grown in carbon-limited continuous culture with either nitrilotriacetic acid (●) with glucose/ammonium (○) as the sources of carbon, nitrogen and energy, as a function of the dilution (growth) rate. Data from Wilberg (1990).

There is considerable information available in

Table 3. Detection and enumeration of NTA-degrading microorganisms in samples collected from different ecosystems and wastewater treatment plants.

Origin of sample	AODC[a]	IFT-positive[b] or MPN[c] (in % of AODC)	Reference
Activate sludge from treatment plant exhibiting good NTA elimination (>98%)	$1.1 \cdot 10^9$	0.026%	Wilberg (1989)
Activated sludge from treatment plant exhibiting reduced NTA elimination (41%)	$2.4 \cdot 10^9$	0.192%	Wilberg (1989)
Kriesbach, small river eutrophic	$1.3 \cdot 10^9$	0.014%	Wilberg (1989)
River Rhine, surface water	$1.4 \cdot 10^9$	0.062%	Wilberg (1989)
Garden soil	$1.1 \cdot 10^8$	0.003%	Wilberg (1989)
Estuary, 0.8–1.9‰ salinity	$5 \cdot 10^8$–$9.2 \cdot 10^9$	0.0002–0.26%	Larson & Ventullo (1986)
Estuary, 1.4‰ salinity	$0.3 \cdot 10^9$	0.22%	Pfaender et al. (1985)

[a] Acridine orange direct counts.
[b] Immunofluorescence test positive cells with antibodies raised against strain ATCC 29600.
[c] Most probable number.

the literature suggesting the necessity for an adaptation period in order to obtain NTA degradation in environmental samples. However, such studies do not answer the question as to whether this adaptation period was due either to enrichment of competent microorganisms or to induction of NTA-degrading enzymes in microorganisms originally present in these samples. Only detailed studies in which microscopic techniques are combined with serological or radiolabelling methods, or the use of gene probes can yield information on the presence and expression of enzymes involved in the degradation of NTA under such complex conditions. Another approach to study the regulation of the capacity to degrade NTA would be the exposure of pure cultures of NTA-degrading bacteria in membrane chambers to the environment. Using this method, McFeters et al. (1990) were able to demonstrate that strain ATCC 29600, pre-grown in the laboratory under non-inducing conditions, was able to induce the catabolic enzymes for NTA within 6–8 hours upon transfer to a wastewater treatment plant.

Other aminopolycarboxylic acids

Presently, no report on the successful isolation and cultivation of either a microbial enrichment culture or a pure culture able to degrade EDTA, DTPA, or HEDTA is available in the open literature. Nevertheless, either biologically mediated transformation or degradation of all three compounds has been reported to occur. Whereas a number of studies dealing with the degradation of EDTA have been published the authors are aware of only one report concerning the degradation of DTPA and HEDTA in soil slurries (Means et al. 1980).

The biodegradation of EDTA is very important with respect to environmental considerations because it is a compound which is currently used in quantities equal to or higher than NTA (Table 1). Biologically-mediated, aerobic degradation of EDTA has been reported in soils, river and lake sediments (Tiedje 1975 and 1977; Means et al. 1980) and in sludge from an aerated wastewater treatment lagoon (Belly et al. 1975). Using ^{14}C-labelled EDTA, Belly and co-workers demonstrated the microbial breakdown to ^{14}C-CO_2 of both the ethylene and the acetate part of the molecule. Because the addition of easily degradable carbon sources stimulated oxidation of EDTA, it was suggested that EDTA was co-metabolized by certain members of the microbial population rather than utilized as a sole source of carbon or nitrogen by a specific EDTA-degrading bacterium (Tiedje 1975 and 1977). In contrast to soil where no significant accumulation of breakdown intermediates could be detected (Tiedje 1975) a range of compounds, probably derived from microbial degradation of EDTA were identified by Belly et al. (1975) in wastewater. Primarily IDA and ED3A and also glycine and NTA were found as products. This and the fact that similar intermediates were found during aerobic photodegradation of Fe(III)-EDTA (Lockhart & Blakeley 1975) points to the possible involvement of light and NTA-degrading microorganisms in the breakdown of EDTA. Despite this, EDTA seems to be rather persistent in wastewater treatment plants (Madsen & Alexander 1985; Bunch & Ettinger 1962; Gerike & Fischer 1979). This is supported by the recent results obtained during the Swiss NTA/EDTA monitoring programme where no significant degradation of EDTA (in contrast to NTA) was detected during aerobic wastewater treatment and in aerobic groundwater infiltration zones. Despite the greatly increased use of NTA in laundry detergents since the Swiss ban of STP in July 1986, a higher residual concentration of EDTA (10–20 $\mu g\, l^{-1}$) than of NTA (2–5 $\mu g\, l^{-1}$) was found in larger Swiss rivers (Giger et al. 1987). Similar observations have been made for the River Ruhr in Germany (Dietz 1987).

Polyphosphates

Polyphosphates are relatively stable compounds that chemically hydrolyse only very slowly. Half-lives in sterile water of the two polyphosphates used as detergent builders, STP and SPP, at pH 7.0 and 30° C were reported to be in the range of 1–3 years (Griffith et al. 1973; Zinder et al. 1981). The biologically catalyzed rate of STP hydrolysis in both natural waters and wastewater treatment plants is two to three orders of magnitude faster (Heinke et al. 1969).

A wide range of eukaryotic and prokaryotic cells are known to accumulate and store high molecular weight polyphosphates within their cells. Under certain conditions bacteria and yeasts have been reported to accumulate as much as 20% of their dry weight as polyphosphate (Kulaev & Vagabov 1983). Hence, it is not surprising that many microorganisms are equipped with the ability to hydrolyse polyphosphates and use them as a source of phosphorus.

With respect to the transport of polyphosphates into cells relatively few data have been published and the basic question of whether polyphosphates can be transported into the cell at all still awaits clarification. According to Kulaev & Vagabov (1983) tripolyphosphate (TPP) seems to be the form in which polyphosphate is taken up most efficiently by cells. Nevertheless, because of numerous reports about extracellular and periplasmic polyphosphate-degrading phosphatases (Kulaev & Vagabov 1983), it seems justified to assume that extracellular polyphosphates are first hydrolysed to inorganic phosphate before being transported into the cell. Considerable evidence for such a mechanism was recently found for *Escherichia coli* (Rao et al. 1987). In this bacterium growth on polyphosphate (chain length approximately 100 phosphate residues) as a sole source of phosphorus was dependent on the presence of alkaline phosphatase. Mutants which were not able to either induce the phosphate regulon or were deficient in an active alkaline phosphatase lost the capability to grow on polyphosphate (Rao et al. 1987; Torriani-Gorini 1987).

Until now several different enzyme types (eq. I–III) hydrolyzing polyphosphates (including TPP) have been described in the literature, most of them exhibiting low specificity with respect to the degree of polymerization of the polyphosphate accepted as the substrate. As discussed by Kulaev (1979) it is not yet known whether this lack of specificity is a general property of these enzymes or whether it originates from the problem of separating individual phosphatases with differing substrate specificities during purification.

I Alkaline and acid phosphatases
$R\text{-}P + H_2O \rightarrow ROH + P_i$

IIa Polyphosphate-phosphohydrolases
$PP_n + H_2O \rightarrow PP_{n-1} + P_i$

IIb $TPP + H_2O \rightarrow PP + P_i$

III Polyphosphate-depolymerases
$PP_n + H_2O \rightarrow PP_{n-x} + PP_x$

In the second group there are specific tripolyphosphatases as they have been described for many fungi, yeasts and bacteria (Kulaev & Vagabov 1983). In eucaryotic cells these tripolyphosphatases were found mainly in the mitochondrial fraction and therefore might be of minor importance in the hydrolysis of extracellular TPP, whereas the localization of tripolyphosphatase in bacteria is still obscure. In general, in both eukaryotes and prokaryotes polyphosphatases are known to be bound to the outside of the cytoplasmic membrane (Kulaev & Vagabov 1983). Their synthesis was reported to be repressible by extracellular phosphate and derepression of these enzymes is usually observed during phosphate starvation (Harold 1966; Torriani-Gorini 1987).

Phosphonates

In earlier reports metal-complexing phosphonic acids were generally considered to be rather stable compounds towards microbial attack. Nevertheless, a range of microorganisms is known to be able to split the C-P bond and remove the phosphorus group from structurally simple alkyl-phosphonates by means of a specific phosphonatase (La Nauze et al. 1978).

Only recently, isolation of microorganisms able to degrade HEDP, EDTMP and ATMP has been described in the literature. Pipke et al. (1987) reported the isolation of an *Arthrobacter* sp. (strain GLP-1) that was able to grow with the phosphonate herbicide glyphosate as the only source of phosphorus. Subsequent investigations showed that this strain was able to utilize a wide range of phosphonic acids when they were supplied as the only source of phosphorus, including the complexing agents HEDP, EDTMP and DTPA during growth with which doubling times in the range of 12–15 hours were measured (Pipke & Amrhein 1988). This was confirmed by Schowanek & Verstraete (1990a) who tested several bacteria with respect to their potential to utilize a range of phosphonates. Whereas most strains were able to obtain phosphorus for growth from structurally simple mono-phosphonates only *Arthrobacter* sp. GLP-1 was able to use the four complexing phosphonates HEDP, EDTMP, ATMP and DTPMP as the sole source of phosphorus. The same authors demonstrated enrichment and isolation of three pure cultures of EDTMP-, DTPMP- and HEDP-degrading bacteria from seven different ecosystems that had been previously exposed to these phosphonates suggesting that microbes capable of degrading these chelating agents are widely distributed in polluted natural waters. One of the isolates was a Gram-negative, non-motile rod (tentatively identified as *Pseudomonas paucimobilis*) which grew in batch culture with HEDP as the sole source of phosphorus with a doubling time of 9.4 hours (Schowanek & Verstraete 1990b). Experiments with *Arthrobacter* sp. GLP-1 indicated that this strain was not able to assimilate nitrogen from these complexing phosphonates under nitrogen/phosphorus-limited growth conditions. This suggests that only the phosphonate group(s) in these compounds was removed and utilized as a growth substrate (Schowanek & Verstraete 1990a). Such action would be consistent with the observations made by Daughton et al. (1979a) who reported that during growth of *Pseudomonas testosteroni* with several alkylphosphonates the alkyl groups were released into the growth medium in the reduced form. One might speculate therefore that the microbial attack of ATMP and HEDP via a similar mechanism would yield methylamine and ethanol, respectively.

With respect to the regulation of phosphonate metabolism, it has been demonstrated for a variety of alkylphosphonate-degrading Gram-negative and Gram-positive bacteria that both transport of phosphonates and phosphonatase are inhibited/repressed in the presence of ortho-phosphate (Rosenberg & La Nauze 1967; Daughton et al. 1979b; Lerbs et al. 1990). Evidence that this pattern of regulation might not hold for all organisms was recently presented by Schowanek & Verstraete (1990b) who found that methylphosphonate was used simultaneously with ortho-phosphate during batch growth of their isolate MMM101a.

Polymeric polycarboxylates

Although these compounds are no chelating agents in the strict sense they are frequently included in today's detergents because of their ability to prevent growth and deposition of crystals on textile fibres (Opgenorth 1987). To date, isolation of a microbial culture able to degrade either PA or PHC has not been reported. In closed-bottle tests degradation of such compounds has proved to be slow and highly dependent on the degree of polymerisation (Haschke & Marlock 1974; Metzner & Nägerl 1982). Studies performed in treatment plants indicated that these polymers are not mineralized during wastewater treatment but are eliminated by either sorption to activated sludge or by precipitation as calcium salts (Opgenorth 1987).

Acknowledgements

The authors are indepted C.A. Mason for his help during the preparation of this manuscript and to D. Schowanek and W. Verstraete for sending us their manuscript in advance of publication.

References

Bunch RL & Ettinger MB (1967) Biodegradability of potential organic substitutes for phosphates. Proc. Purdue University International Waste Conference. Engineering Bulletin of the Purdue University, Engineering Extension Series 129: 393–396

Belly RT, Lauff JJ & Goodhue CT (1975) Degradation of

ethylenediaminetetraacetic acid by microbial populations from an aerated lagoon. Appl. Microbiol. 29: 787–794

Berth P, Berg M & Hachmann K (1983) Mehrkomponentensysteme als Waschmittelbuilder. Tenside and Detergents 20: 276–282

Bernhardt H (1990) An ecological assessment of organic phosphate substitutes. Vom Wasser 74: 159–176

Cripps RE & Noble AS (1973) The metabolism of nitrilotriacetate by a Pseudomonad. Biochem. J. 136: 1059–1068

Daughton CG, Cook AM & Alexander M (1979a) Bacterial conversion of alkylphosphonates to natural products via carbon-phosphorus bond cleavage. J. Agric. Food Chem. 27: 1375–1382

Daughton CG, Cook AM & Alexander M (1979b) Phosphate and soil binding: factors limiting bacterial degradation of ionic phosphorus-containing pesticides. Appl. Environ. Microbiol. 37: 605–609

De Vos P & De Ley J (1983) Intra- and intergeneric similarities of *Pseudomonas* and *Xanthomonas* ribosomal ribonucleic acid cistrons. Int. J. Syst. Bacteriol. 33: 487–509

Dietz F (1987) Neue Messergebnisse über die Belastung von Trinkwasser mit EDTA. Wasser Abwasser gwf 128: 286–288

Egli T (1988) (An)aerobic breakdown of chelating agents used in household detergents. Microbiol. Sci. 5: 36–41

Egli T, Weilenmann H-U, El-Banna T & Auling G (1988) Gram-negative, aerobic, nitrilotriacetate-utilizing bacteria from wastewater and soil. Syst. Appl. Microbiol. 10: 297–305

El-Banna T (1989) Characterization of some unclassified *Pseudomonas* species. PhD thesis, Tanta University, Egypt

Epstein SS (1972) Toxicological and environmental implications of the use of nitrilotriacetic acid as a detergent builder. Int. J. Environ. Stud. 2: 291–300

Firestone MK & Tiedje JM (1978) Pathway of degradation of nitrilotriacetate by a *Pseudomonas* species. Appl. Environ. Microbiol. 35: 955–961

Firestone MK, Aust SD & Tiedje JM (1978) A nitrilotriacetic acid monooxygenase with conditional NADH-oxidase activity. Arch. Biochem. Biophys. 190: 617–623

Focht DD & Joseph HA (1971) Bacterial degradation of nitrilotriacetic acid. Can. J. Microbiol. 17: 1553–1556

Gerike P & Fischer WK (1979) A correlation study of biodegradability determinations with various chemicals in various tests. Ecotoxicol. Environ. Safety 3: 159–173

Giger W, Ponusz H, Alder A, Baschnagel D, Renggli D & Schaffner C (1987) Auftreten und Verhalten von NTA und EDTA in schweizerischen Gewässern und im Trinkwasser. EAWAG Jahresbericht, pp 9–12

Griffith EJ, Beeton A, Spencer JM & Mitchell DT (1973) Environmental Phosphorus Handbook. Wiley, New York

Haschke H & Marlock G (1974) Untersuchungen über mikrobiologisch-physiologisch relevante Eigenschaften bestimmter Poly(hydroxycarboxylat)-Komplexbildner (1. Mitteilung). Tenside and Detergents 11: 57–74

Harold FM (1966) Inorganic polyphosphates in biology: Structure, metabolism, and function. Bacteriol. Rev. 30: 772–794

Hauptausschuss 'Phosphate und Wasser' (1978) Phosphor – Wege und Verbleib in der Bundesrepublik Deutschland. Verlag Chemie, Weinheim

Heinke GW (1969) Hydrolysis of condensed phosphates in Great Lakes Waters. Proc. 12th Conference on the Great Lakes Research

Ikemoto S, Suzuki K, Kaneko T & Komagata K (1980) Characterization of strains of *Pseudomonas maltophila* which do not require methionine. Int. J. Syst. Bacteriol. 30: 437–447

Jakobi G, Lühr A, Schwuger MJ, Jung D, Fischer W & Gloxhuber C (1983) Waschmittel. In: Ullmanns Encyklopädie der technischen Chemie (4th edition), Vol 24 (pp 63–160). Verlag Chemie, Weinheim

Kakii K, Yamaguchi H, Iguchi Y, Teshima M, Shirakashi T & Kuriyama M (1986) Isolation and growth characteristics of nitrilotriacetate-degrading bacteria. J. Ferment. Technol. 64: 103–108

Kemmler J & Egli T (1990) Nitrilotriacetat-abbauende Mikroorganismen. Wasser Abwasser gwf 131: 251–255

Kemper HC, Martens RJ, Nooi JR & Stubbs CE (1975) Nitrogen- and phosphorus-free strong sequestering builders. Tenside and Detergents 12: 47–51

Kulaev IS (1979) The Biochemistry of Inorganic Polyphosphates. Wiley, Chichester

Kulaev IS & Vagabov VM (1983) Polyphosphate metabolism in micro-organisms. Adv. Microb. Physiol. 24: 83–171

La Nauze JM, Rosenberg H & Shaw DC (1970) The enzymic cleavage of the carbon phosphorus bond: purification and properties of phosphonatase. Biochim. Biophys. Acta 212: 332–350

Larson R & Ventullo R (1986) Kinetics of biodegradation of NTA in an estuarine environment. Ecotoxicol. Environ. Safety 12: 166–179

Lerbs W, Stock M & Parthier B (1990) Physiological aspects of glyphosate degradation in *Alcaligenes* sp. strain GL. Arch. Microbiol. 153: 146–150

Lockhart HB & Blakeley RV (1975) Aerobic photodegradation of Fe(III)-(Ethylenedinitrilo)tetraacetate(ferric EDTA). Environ. Sci. Technol. 9: 1035–1038

Madson EL & Alexander M (1985) Effects of chemical speciation on the mineralization of organic compounds by microorganisms. Appl. Environ. Microbiol. 50: 342–349

Metzner EA, Crutchfield MM, Langguth RP & Swisher RD (1973) Organic builder salts as replacements for sodium tripolyphosphate (II). Tenside and Detergents 10: 239–245

Metzner G & Nägerl HD (1982) Environmental behaviour of two water-conditioning agents based on phosphonate and polyacrylate. Tenside and Detergents 19: 23–29

McFeters GA, Egli T, Wilberg E, Alder A, Schneider R, Snozzi M & Giger W (1990) Activity and adaptation of nitrilotriacetate (NTA)-degrading bacteria: field and laboratory studies. Water Res. 24: 875–881

Means JL, Kucak T & Crerar DA (1980) Relative degradation rates of NTA, EDTA and DTPA and environmental implications. Environ. Pollut. Ser. B (Series B) 1: 45–60

Mottola HA (1974) Nitrilotriacetic acid as a chelating agent: applications, toxicology and bioenvironmental impact. Toxicol. Environ. Chem. 44: 81–113

Opgenorth HJ (1987) Umweltverträglichkeit von Polycarboxylaten. Tenside and Detergents 24: 366–369

Pfaender FK, Shimp RJ & Larson RJ (1975) Adaptation of estuarine ecosystems to the biodegradation of nitrilotriacetic acid: effects of preexposure. Environ. Toxicol. Chem. 4: 587–593

Pickaver AH (1976) The production of N-nitrosoiminodiacetate from nitrilotriacetate and nitrate by microorganisms growing in mixed culture. Soil Biol. Biochem. 8: 13–17

Pipke R, Amrhein N, Jacob GS, Schaefer J & Kishore GM (1987) Metabolism of glyphosate in an *Arthrobacter* sp. GLP-1. Eur. J. Biochem 165: 267–273

Pipke R & Amrhein N (1988) Degradation of the phosphonate herbicide glyphosate by *Arthrobacter atrocyaneus* ATCC 13752. Appl. Environ. Microbiol. 54: 1293–1296

Rao NN, Roberts MF & Torriani A (1987) Polyphosphate accumulation and metabolism in *Escherichia coli*. In: Torriani-Gorini A, Rothman FG, Silver S, Wright A & Yagil E (Eds) Phosphate Metabolism and Cellular Regulation in Microorganisms (pp 213–219). American Society of Microbiology, Washington

Rosenberg H & La Nauze JM (1967) The metabolism of phosphonates by microorganisms. The transport of aminoethylphosphonic acid in *Bacillus cereus*. Biochim. Biophys. Acta 141: 79–90

Schneider R (1984) Weltproduktion und -verbrauch an Seifen, Wasch- und Reinigungsmitteln 1980, 1981 and 1982. Tenside and Detergents 21: 212–215

Schneider RS (1989) The NTA-monooxygenase from *Pseudomonas* sp. ATCC 29600. PhD thesis, ETH-Nr. 8824, Zürich

Schowanek D & Verstraete W (1990a) Phosphonate utilization by bacterial cultures and enrichments from environmental samples. Appl. Environ. Microbiol. 56: 895–903

Schowanek D & Verstraete W (1990b) Phosphonate utilization by bacteria in the presence of alternative phosphorus sources. Biodegradation 1: 43–53

Shannon JE & Lee GF (1966) Hydrolysis of condensed phosphates in natural waters. Air Water Pollut Int. J. 10: 735–756

Snozzi M & Egli T (1987) Purification of a NTA-monooxygenase. Proc. 4th European Congress on Biotechnology 3: 345

Tiedje JM (1975) Microbial degradation of ethylenediaminetetraacetate in soils and sediments. Appl. Microbiol. 30: 327–329

Tiedje JM (1977) Influence of environmental parameters on EDTA biodegradation in soils and sediments. J. Environ. Qual. 6: 21–26

Tiedje JM (1980) Nitrilotriacetate: Hindsight and gunsight. In: Maki AW, Dickson KL & Cairns J (Eds) Biotransformation and Fate of Chemicals in the Aquatic Environment (pp 114–119). ASM, Washington

Tiedje JM, Mason BB, Warren CB & Malek EJ (1973) Metabolism of nitrilotriacetate by cells of *Pseudomonas* species. Appl. Microbiol. 25: 811–818

Torriani-Gorini A (1987) The birth and growth of the Pho regulon. In: Torriani-Gorini A, Rothman FG, Silver S, Wright A & Yagil E (Eds) Phosphate Metabolism and Cellular Regulation in Microorganisms (pp 3–11). ASM, Washington

Vollenweider R (1968) Die wissenschaftlichen Grundlagen der Seen- und Fliesswassereutrophierung unter besonderer Berücksichtigung der Phosphors und des Stickstoffs als Eutrophierungsfaktoren. OECD Report, Paris

Wanner U, Egli T & Snozzi M (1989) A dehydrogenase as the first step in the anaerobic pathway for nitrilotriacetate (NTA) degradation. In: Hamer G, Egli T & Snozzi M (Eds) Mixed and Multiple Substrates and Feedstocks (pp 165–167). Hartung-Gorre, Constance

Wanner U, Kemmler J, Weilenmann HU, Egli T, El-Banna T & Auling G (1990) Isolation and growth of a bacterium able to degrade nitrilotriacetic acid under denitrifying conditions. Biodegradation 1: 31–42

Wehrli E & Egli T (1988) Morphology of nitrilotriacetate-utilizing bacteria. Syst. Appl. Microbiol. 10: 306–312

Wilberg E (1990) Zur Physiologie und Ökologie Nitrilotriacetat (NTA) abbauender Bakterien. PhD thesis, ETH-Nr. 9015, Zürich

Wong PTS, Liu D & McGirr DJ (1973) Mechanism of NTA degradation by a bacterial mutant. Water Res 7: 1367–1374

Zinder B, Hertz J & Oswald HR (1984) Kinetic studies on the hydrolysis of sodium tripolyphosphate in sterile solution. Water Res 18: 509–512

Biodegradation **1**: 133–146, 1990.

Physiology and performance of thermophilic microorganisms in sewage sludge treatment processes

B. Sonnleitner & M. Bomio
Department for Biotechnology, Swiss Federal Institute of Technology, ETH Zürich Hönggerberg, CH 8093 Zürich, Switzerland

Introduction

A combined treatment of domestic and industrial liquid wastes is preferably used in most industrialized countries. The primary objective of wastewater treatment plants is a high degree of water purification. But relatively little attention is payed to the primary byproduct, sewage sludge. The economical value of sewage sludge is among the lowest of all materials available on earth, but sludge will undoubtedly continue to be produced in very large quantities in the future. Its effective treatment will be one of the most important problems to solve.

In Switzerland, 95% of the population are connected to one of the more than 1000 wastewater treatment plants. The production of wet sewage sludge is 4′000′000 m^3 a^{-1} or 250′000 t of dry matter a^{-1}, corresponding to 0.1 kg of dry matter d^{-1} $person^{-1}$. This high specific quantity of sewage sludge is typical for industrialized countries and is expected to increase in the future, particularly in those countries where wastewater treatment is still expanding (personal communications BUWAL (= Environmental Protection Agency of CH) 1990).

The sludge produced in Switzerland is presently used in agriculture (as a substitute for synthetic fertilizers), deposited in landfill sites, or incinerated. The amount recycled in agriculture is presently 30 to 50% of the total production; the percentage varies from region to region. About 40 to 50% of the sludge produced are deposited in landfills and 10 to 30% are incinerated.

The use of sewage sludge in agriculture is declining: at the beginning of the 1980s, the amount used was 50% of the total sludge production. Realistic estimates predict that this quantity will reduce to 20% or 50′000 of dry matter t a^{-1} in the next years (Schweizer 1988). Extended or prolonged use of sewage sludge as fertilizer is possible only if waste sludge can be processed in a way to render appropriate bacteriological, chemical and physical properties.

Biological processes for the treatment of the sewage sludge have been implemented in 75% of all wastewater treatment plants in Switzerland. Until now, the most popular biological treatment techniques established were either mesophilic anaerobic stabilization in large plants or *aerobic* cold stabilization (uncontrolled process: just aeration, no elevation of temperature) in small plants. The *aerobic thermophilic* process represents a relatively new technology: it can easily be integrated in the traditional treatment sequence or it can substitute for the anaerobic process. In any way, the so treated sewage sludge becomes hygienized and its physical chemical quality is improved. Table 1 summarizes the types of sludge bioprocessing used in Switzerland (BUWAL 1990). The 20 plants exploiting an aerobic thermophilic sludge treatment step (ATS) constructed in Switzerland during the 1980s do not reflect just improved microbiological understanding or technological innovation, on the contrary, this was caused by legislation (Klärschlammverordnung, Federal Department of Interior 1981) prescribing that 100 *Enterobacteriaceae* per g of sludge must not be exceeded for use in agriculture. Further, this sludge must also be free of parasitic worm eggs and viruses.

Between 1975 and 1985, the development of the

aerobic thermophilic sludge treatment technology was characterized by rapid progression to practice without the development of a sound basic scientific background (i.e.: trial and error instead of deterministic design based on knowledge of physiology and/or population dynamics). The majority of the studies carried out improved performance of sludge treatment in existing plants. However, virtually no attempt was made to examine the physiology of the microorganisms involved in aerobic thermophilic treatment using the only realistic substrate – sewage sludge itself.

A biotechnological approach to the study of sludge treatment involves investigations of microorganisms as biocatalysts in the process environment. However, the currently proposed process for sludge biotreatment differs from most biotechnological processes in the following ways:

- application of mixed populations (non sterile)
- complex and very ill-defined substrates
- largely varying substrate feeds (qualitatively and quantitatively)
- irrelevant commercial value of the final product
- non-profit basis for process operation
- low technology

Aerobic bioprocesses for the treatment of sewage sludge at high temperatures offer many advantages. The benefits of the technology allow the product to be safely used as an agricultural fertilizer. When disposal is the favoured option, the quality of the treated sludge can definitely be improved. Advantages are summarized as follows:

- biodegradation of normally biodegradable organic components at high rates
- biodegradation of recalcitrant compounds
- improvement of the physical characteristics of the sludge
- stability and reliability of the process due to the uniform microflora employed (selectivity pressure of the temperature)
- sanitization of the sludge (inactivation of all potentially pathogenic organisms and viruses)

The sanitization of sludge can also be achieved using non-biological thermal technology, by physical processes such as irradiation or chemical procedures such as the addition of lime. However, such processes do not incorporate the full range of benefits that can be gained with bioprocessing.

The treatment of the sewage sludge with thermophilic microorganisms represents an example of a low technology bioprocess (Bergquist et al. 1987). The non-profit nature of wastewater treatment industry generally inhibits innovation, but legislation can give the necessary positive impulse to improve the *status quo*.

Definition and choice of appropriate objectives for a process must be the starting point for all experimental work. This should be done on the basis of both theoretical and practical considerations, e.g. by evaluating different scenarios. It is very important to use such a dual approach in the wastewater treatment industry, where different interest groups interact. A range of different goals possibly influence the development of sludge treatment processes. Chemically, the maximal reduction of organic matter is a primary objective. Physically, dewaterability and heat production are most important goals, but also retention time is a central factor. Consideration of biological factors indicates four different goals: the hygienization of the sludge is the cardinal factor to be achieved in an aerobic thermophilic treatment. At the same time, high biocatalytic activity and low biomass yield are desired. When the aerobic thermophilic treatment is

Table 1. Total number and percentage of wastewater treatment plants with integrated bioprocesses for sewage sludge treatment. Both, producing and projected plants are included reflecting the situation in Switzerland in 1990.

Sludge treatment	Number of plants	Percentage
Anaerobic mesophilic	575	75%
Aerobic cold	139	18%
Sanitization	52 = 23 + 29*	7%
Total	766	100%

Anaerobic mesophilic = classical method to stabilize sludge, T approximately 35° C.
Aerobic cold = uncontrolled process: just aeration, no elevation of temperature.
Sanitization* = ATS treatment (23 plants) and thermal inactivation (29 plants).

considered as a pretreatment, a further goal is an 'improvement' of the substrate sludge for the next process step.

The treatment of sewage sludge with aerobic thermophilic microorganisms is a genuine practical application of thermophilic process microbes. Most development work has been done in Germany, Switzerland, Sweden, UK, Republic of South Africa and USA. Unfortunately, quantitative data about physiology, kinetics and population dynamics that would permit process development and design are generally missing.

Methodology

Several different bioreactors were used in our own investigations: optimization works were carried out in a compact loop bioreactor of 7 l total volume. Studies on inactivation of pathogens and experiments exploiting the fed batch cultivation technique were carried out in a 50 l compact loop reactor (COLOR): these reactors were equipped with mechanical foam breakers (Hess 1988). A pilot scale bioreactor (4000 l total volume) was a commercial design but modified during the experimental phase. The characteristic properties of these reactors are summarized in Table 2. The pilot scale reactor is shown in Fig. 1.

During all experiments, the following variables were measured on-line: fluxes of liquid and gas phases, temperature, pH, pO_2, pCO_2, redox potential, and exhaust gas composition. To our knowledge, this is the first time that an extensive on line measurement of sewage sludge populations has been performed. Detailed description of microorganisms, media, off-line analyses, and sewage sludge can be found elsewhere (Bomio 1990).

Results and discussion

Thermophilic populations and their dynamics

The characterization of representative sets of process microorganisms from aerobic thermophilic sludge treatment plants according to their microbiological and biochemical properties resulted in the classification of such isolates as *Bacillus stearothermophilus* populations. This was based on comparisons with references strains and held true for at least 95% of all strains, while 5% of the strains could not be safely identified as *Bacillus* only because sporulation could not be induced. All isolates showed extremely high specific growth rates (μ_{max} in the order of $2\,h^{-1}$) and generally low yield coefficients ($Y_{x/s}$ between 0.2 and $0.3\,g\,g^{-1}$; Sonnleitner & Fiechter 1983c).

These thermophilic populations respond with a fast reaction to changes of the substrate composition when cultivated in defined media. The lag phases after different fluctuations in substrates composition vary between 1 and 5 h, depending on the nature of the new substrate. A similar response was observed in pilot scale: high amylase activity (with an optimum at 70° C) was observed to appear rapidly during experiments, when the feed sludge contained some starch (e.g. potato residues from

Table 2. Characteristics of reactors used for studies on lab and pilot scale according to Bomio 1990 (COLOR = compact loop reactor).

		7 l Color	50 l Color	Pilot scale reactor
Total volume	[m³]	0.007	0.050	4
Working volume	[m³]	0.004	0.014	3
Height	[m]	0.030	0.792	2.68
Diameter	[m]	0.015	0.261	1.38
Max. aeration	[vvm]	1.5	1.5	0.1
Power input		impeller	impeller	circulation pump (loop)
Heat exchange		reactor jacket	reactor jacket	external loop
Foam breaker		external	internal	internal
Insulation		none	none	10 cm Alufloc

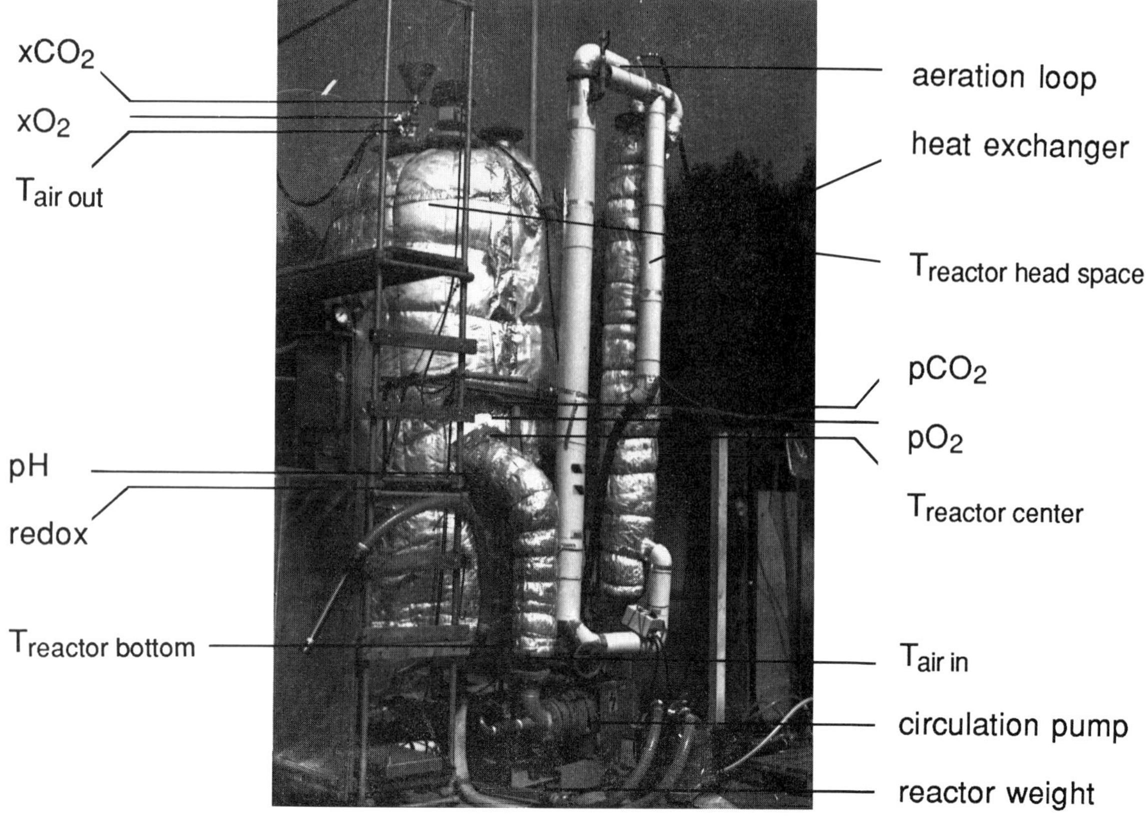

Fig. 1. The pilot scale bioreactor and measurement equipment: several temperature (T) sensors, sensors for partial pressures (pO_2 and pCO_2), redox potential and exhaust gas composition (xO_2 and xCO_2).

food industry; Bomio, 1990). Although the presence of starch in raw sludge is rare, these results show that the thermophilic populations are able to react immediately and efficiently to changing (chemical) environmental conditions, as illustrated in Fig. 2.

A similar rapid dynamic response of the process microflora was reported upon glucose addition to raw sewage sludge during continuous cultivation: the production of acetic and other volatile fatty acids was found to be induced without delay (Baier 1987). This situation is generally not observed in raw sewage sludge: either oxygen limited or unlimited sludge treatment processes at different scales show that the acidification of the sewage sludge is very unlikely (i.e. the exception rather than the rule).

Baier et al. (1986) have compiled some other characteristic population dynamic aspects of isolated thermophilic process bacteria grown in the lab over many hundreds of generations. The following pheno- and genotypic variations of the original strains were found significant:

Observed difference	Origin/effect
genetics	loss of plasmids
physiology	restricted exoenzyme formation
kinetics	different kinetic behavior
morphology	altered cell wall structure
phenotype	cell shape, culture appearance
metabolic control	several inhibition effects

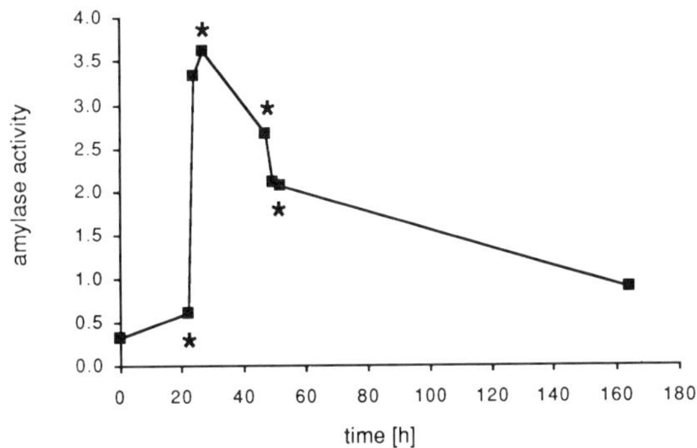

Fig. 2. Time course for amylase activity during fed batch cultivations in the 4'000 l pilot scale bioreactor. The mean hydraulic retention time was 14 h, volume changes every 2.5 h (*). Starch was present in the lot of sludge replaced at time 20 h, and amylase activity increased rapidly showing the short adaptation time of the thermophiles to new environmental conditions. Following volume changes after activity maximum showed a characteristic wash out behavior: the subsequent lots of sludge changed did no longer contain starch.

Biocatalytic activities

Biomass

Microbial growth can be easily determined in clear culture media by measuring the increase of cell mass (biomass). This is essential for the evaluation of both biomass yield coefficients and specific growth rates. But in sewage sludge, which contains particulate matter, this direct method cannot be applied. Hence, it is necessary to exploit any indirect methodology that can be correlated with biomass formation or activity. Several methods are available for the determination of biomass on the basis of a chemical index (White et al. 1979). Indirect variables considered and the evaluation of five methods are summarized in Table 3. Of greatest interest for evaluation are the applicability in real sewage sludge, good correlation with biomass concentration, reproducibility and simplicity of the analytical procedure.

The determination of ATP proposed as biomass indicator (Holm-Hansen 1973; Droste & Sanchez 1983) is not found satisfactory: instability of the ATP extract and pronounced variations of ATP concentrations depending on growth phase and substrate being metabolized (King & White 1977) are the restricting factors.

The cell plate count method, as proposed for mesophilic aerobic sludge samples (Droste & Sanchez 1983), can give satisfactory results but it is laborious, very time consuming and difficult to apply for (mixed) thermophilic populations. The general disadvantage of the method is the inherent underestimation of the 'true' cell number concentration. The existence of cell aggregates in sludge causes a misinterpretation of raw data because the assumption – a colony derives from 1 single cell – is obviously not true.

The overall dehydrogenase activity reflects the rate of oxygen consumption under strictly aerobic metabolic conditions (Lopez et al. 1986). After method optimization, dehydrogenase activity is found to be proportional to biomass concentration when the oxidative metabolism is not limited.

DNA is used to estimate microbial biomass in activated sludge processes (Thomanetz 1982). The determination of DNA as a biomass indicator is excellent for samples from defined media cultures but extremely unpracticable from sludge when interference from the complex matrix become dominant.

Phospholipids, also an indirect estimate for microbial mass, shows two important restrictive factors when applied to determine growth of aerobic thermophilic process organisms. These are the presence of high concentrations of lipids in the raw sludge and the metabolism of the phospholipids by thermophilic organisms. Therefore, the method was not found very successful.

The evaluation for following microbial growth in/on sewage sludge is: total dehydrogenase activity can be most reliably used as an indirect measure for biomass. However, it reflects the cells activity rather than their mass concentration.

Limiting factors

Growth is generally under the control of extracellular chemical factors such as available carbon energy source(s), nutrients, trace elements and growth factors. Sewage sludge is both, ill-defined and highly variable with respect to concentration and availability. Experiments designed to identify the dynamics of suitable carbon sources and limiting factors for thermophilic bacteria during the treatment process, however, suffer from analytical difficulties.

The nature of limitations responsible for the decrease of both, the oxygen uptake rate and the total dehydrogenase activity during limited growth were determined by pulsing different carbon energy substrates, salts, trace elements and vitamins to the cultures. Figure 3 shows a typical response to a carbon substrate pulse: the increase of the oxygen uptake rate and the simultaneous decrease of the oxygen partial pressure in the sludge immediately after pulsing fructose shows that a carbon limitation had been released.

The pulse experiments done at the end of fed batch cultures clearly show that biodegradable carbon energy source availability is the only limitation there and salts, trace elements and vitamins are not limiting. Neither is oxygen limiting because pO_2 was always $\geq 80\%$ of air saturation during these pulse experiments (lab scale studies with raw sludge, Bomio et al. 1989).

Activities in sewage sludge and pure cultures

The above cited experiments indicate permanent proteolytic activity because growth was observed immediately after pulses of either casein, gelatine and soya protein. This could be confirmed by measuring the total activity of proteolytic enzymes. Isolated mixed sub-populations of thermophilic process bacteria grew on semi-synthetic media with soya protein as the sole carbon source (Bomio et al. 1989). Further extended pulse experiments showed that extracellular proteases are the only polymer degrading enzymes produced during growth of the process microflora on sewage sludge. Other non-proteinaceous polymers generally present in sewage sludge represent potential substrates for thermophiles, but are normally not hydrolyzed by the process populations. The activities tested are summarized in Table 4.

The proteolytic activity was found to correlate with the increase of the oxygen uptake rate during both, batch and repetitive batch growth of aerobic thermophilic microorganisms on sewage sludge. This indicates a tightly growth associated production of proteolytic enzymes (Fig. 4) which also correlates with an increase in free ammonium concentration in the culture.

The respiratory quotient (RQ value) in sewage sludge cultivations had an average value of 0.82 during both exponential and oxygen limited growth

Table 3. Comparison of indirect methods for biomass determination.

Method	Applicability in defined media	Applicability in sewage sludge	Validity	Reproducibility	Analytical ease	Overall evaluation
ATP	–	–	–	+	–	–
cell count	++	++	+	–	–	=
dehydrogenase	++	+	+	+	+	++
DNA	++	=	+	+	+	+
phospholipids	+	+	–	+	+	+

++ : very good
\+ : good
= : satisfactory
– : not satisfactory

phase; this is typical for oxidative metabolism of proteins (Lentner 1981). This value then rises to 1.1 in the carbon limited phase indicating changes in the overall metabolism. Control cultivations performed on a semi-synthetic media with the same inoculum but with glucose as sole carbon source showed an average RQ of 1.05 which is typical for fully oxidative growth on carbohydrates.

The above referenced findings (RQ values, proteolytic activity, the pulse experiments with protein substrates and the growth associated production of free ammonium) strongly indicate that the thermophilic populations use proteinaceous material as their preferred carbon source.

Determinations of lipase, cellulase and pectinase activities in ATS processes failed; negative results were obtained with both the original and modified analytical methods, i.e. after specific adaptations to the complex medium sludge.

Cultivations on synthetic medium showed that the proteolytic activity is not cell wall bound: high activity was found in the cell free supernatant. On the other hand, activities determined in the supernatant of cultivations on sewage sludge as substrate were comparably low. The conclusion that the enzymes are excreted into the medium but immediately absorbed by particles in the sludge, is indicated by the behavior shown in Fig. 5: during an initial lag phase, there is no *de novo* formation of proteases. However, in this phase, the sludge contains a significant amount of proteolytic enzymes.

Table 4. Extracellular enzymatic activities in sewage sludge evaluated after pulses of polymers and determination of the respective enzymatic activities.

Enzymes	Increase of OUR after a pulse experiment	Enzymatic activity found
Protease	+	+
Amylase	–	*
Cellulase	–	–
Keratinase	–	–
Lipase	–	–
Pectinase	–	–

+ : always found
– : never found
* : occasionally found

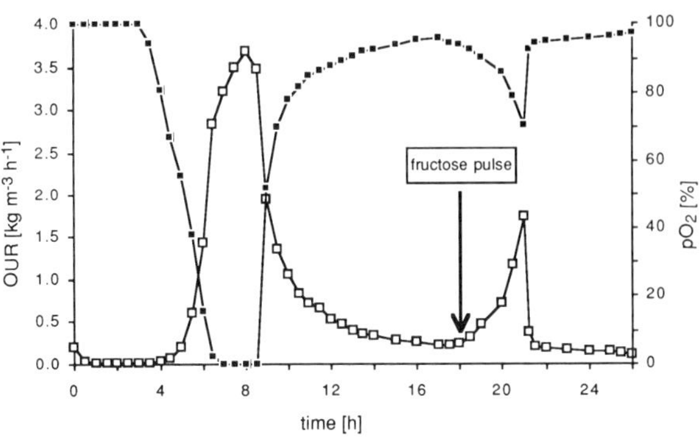

Fig. 3. Oxygen uptake rate (□) and oxygen partial pressure pO_2 (■) during a fed batch cultivation at 65° C, 1500 min^{-1}, 1 vvm and pH 7 in a lab scale reactor: fructose pulse experiment in the carbon limited growth phase. The exponential increase of the OUR immediately after the pulse shows that the fructose pulsed released a (carbon) limitation.

After the lag phase, thermophilic populations begin to grow on sludge with an RQ of 0.81. The proteolytic activity measured in vitro at 80° C and pH 7 rises rapidly, indicating that thermophilic proteases are now formed and also excreted into the medium. These proteases show little activity at 65° C (Sundaram 1988). Activity measured at this temperature does not increase during the growth phase. The hydrolysis of proteinaceous materials with subsequent metabolism of amino acids (obviously oxidative desamination) causes ammonium to be released into the medium. The proteolytic activity originating from the thermophiles has an optimum around 80° C, but at this temperature, raw sludge proteases are not active. Therefore, microbiological activity and process performance can be effectively followed by measuring proteolytic activity at 80° C (Bomio et al. 1989).

Cryptic growth

Some authors have suggested the existence of either antibiotic or 'antagonistic' effects of thermophilic populations on potentially pathogenic microorganisms (Nebiker 1981, Hammel 1983). However, inactivation experiments showed no significant influence of thermophiles on the inactivation of potential pathogens added to raw sludge. Experimental determination of possible antimicrobial substances, using the MIC (Minimal Inhibitory Concentration) test (Hamilton-Miller 1977; Kres-

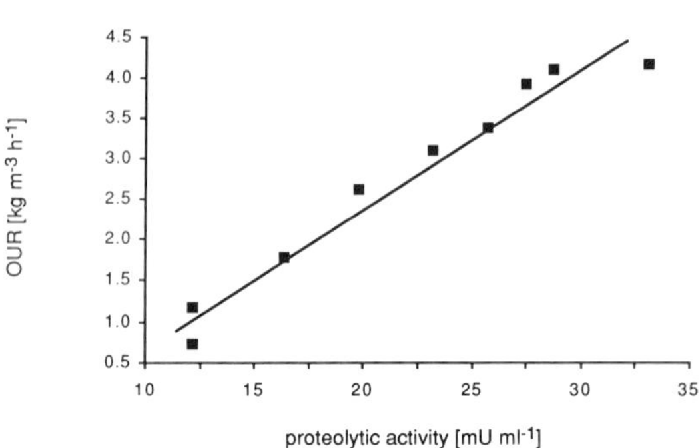

Fig. 4. Linear correlation between proteolytic activity and oxygen uptake rate during a batch cultivation on sewage sludge at 2500 min^{-1}, 65° C, 0.8 vvm and pH 7 in a lab scale reactor. Correlation coefficient r = 0.935. The activity was determined relative to the calibration standard proteinase K (Sigma P-0390, type XI) measured at 80° C. The correlation indicates a growth associated production of extracellular protease by the process microbes.

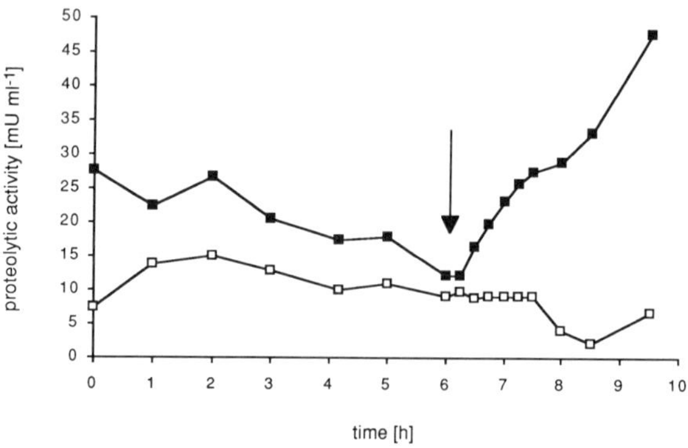

Fig. 5. Proteolytic activity in the sludge (■) and supernatant (□) measured at 80° C and pH 7 during a sludge cultivation at 2500 min^{-1}, 65° C, 0.8 vvm and pH 7 in a lab scale reactor. The activity is given in terms of proteinase K standards. The arrow shows the end of the lag phase. The activity measured in the supernatant does not increase after the lag phase, indicating that the proteolytic enzymes are absorbed to sludge particles.

ken & Wiedmann 1984), conclusively confirmed the absence of such effects.

A non-selectively enriched population of strictly mesophilic microorganisms isolated from sewage sludge, *Escherichia coli* and *Staphylococcus aureus* were cultivated on Isosensitest Broth. Different amounts of raw and aerobic thermophilic (= treated) sludge were added to these cultures. Some results are shown in Fig. 6; similar results were obtained with *E. coli* and *S. aureus*. Growth of the mesophilic microorganisms was enhanced by the addition of sludge (both raw and treated) and not inhibited. The enhancement is linearly correlated with the amount added and significantly greater for aerobic thermophilic than for the raw sludge. This typical response rules out any production of antimicrobial agents during aerobic thermophilic sludge treatment. Its means that some substrates have been made 'more accessible' during thermophilic aerobic treatment of sludge (ATS); even more, the described ATS-process behaves really as a pre-treatment and is not a full stabilization. Comparable results were obtained with thermophilic populations isolated from sewage sludge at temperatures of 45° C (Wassen 1975).

Thermophilic populations isolated from sewage sludge were shown to be able to grow in semi-synthetic media utilizing intact yeast cells as sole substrate (Mason et al. 1986, Hamer and Mason 1987). The formation of a lytic enzyme by *Bacillus stearothermophilus* cultures after induction by mitomycin C has also been described (Walker & Campbell 1966). These observations stimulated the search for lysozyme excreted by mixed populations of *B. stearothermophilus* strains grown on sewage sludge. Unfortunately, lytic activity was found generally to be extremely low in sewage sludge (Bomio et al. 1989). These experiments suggest that inactivation of pathogenic microorganism is a predominantly thermal process and synergistic effects originating from the thermophilic process microbes are negligible.

Inactivation of pathogenic microorganisms

In the early 1970s, inactivation of potentially pathogenic organisms in sewage sludge became an important goal. This resulted in the introduction of different competing technologies. After disastrous experience with pasteurization units installed after the anaerobic digestion step, ATS or pasteurization processes were introduced as pre-treatment subprocesses. But only a few kinetic data for microbially mediated inactivation processes are yet available.

Enterobacteriaceae are widely used as indicator organisms for 'potentially pathogens'. They can be

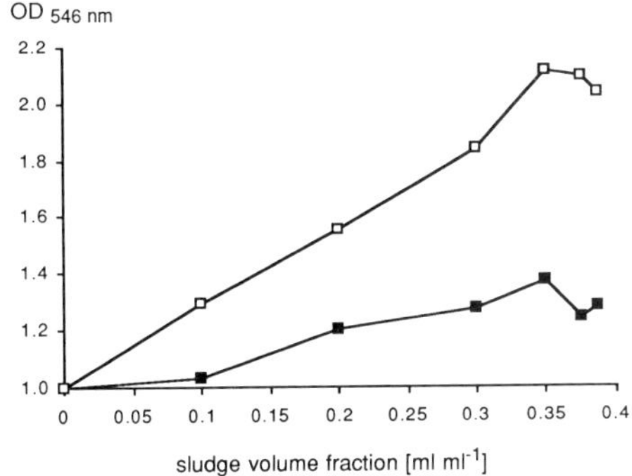

Fig. 6. Relative absorption at 546 nm of a non-selectively enriched culture of mesophiles after 18 h of incubation at 37° C with different fractions of either raw (■) or aerobic thermophilic sludge (□) added to the cultures. The positive and constant slopes obtained indicate that there are no inhibitory effects in both raw and ATS sludge. Treated sludge accelerated the growth of potential pathogens even better (greater slope) demonstrating that this ATS process is operated as a *pre*-treatment step prior to anaerobic digestion and not as a single stabilization process.

selectively determined in homogeneous aqueous suspension after plating on VRBD-agar (Violet Red Bile Dextrose). This classical technique (established in food or environmental analysis) fails when used to analyse sludge samples. Many oxidase positive or slowly growing species occur and make a sound quantitative analysis impossible. This uncertainty has of course forensic consequences; it would be wise to consider the use of other indicator strains which can be easily and reliably quantified for routine hygienic analyses.

Process development

Objective and constraints

Definition of objectives and constraints is one of the most critical points during development of a process for the treatment of sewage sludge. Table 5 summarizes such an evaluation. The evaluation of the variables that are of significance for process performance is crucial. Simultaneous multivariable optimization should be employed because the interdependencies of process variables and parameters are not known a priori for the system sewage sludge (missing background in physiology and population dynamics).

Fed batch cultivations carried out at different temperatures, but with all other variables (pH, stirrer speed, aeration rate and inoculum preparation) kept constant, revealed the effect of temperature on the process performance. This can be monitored by the total oxygen uptake rate, which is proportional to the heat generated by the growing culture (Birou et al. 1987). There was only a minor dependence of the total oxygen uptake rate on temperature found in the range 64 to 68° C. This result confirms well the choice for the process temperature already used in technical-scale plants operating (in Switzerland 65° C). An identical result was obtained for spontaneous thermophilic composting of sewage sludge with poplar sawdust (Viel et al. 1987).

Temperatures lower than 60° C are not considered further because of the decreasing efficiency of sanitization (Feachem et al. 1983; Sonnleitner & Fiechter 1983a, b). Spontaneous population development at temperatures > 70° C occurs only after a considerable lag; at 74° C this extends to 36 hours. This is, of course, inappropriate for a process targeting to auto-generate the necessary heat for the hygienization effect. Growth was observed in processes operated up to 76° C and this result is in agreement with other work on *B. stearothermophilus* (Gibson & Gordon 1975; Baier 1987).

The effect of pH can be examined during cultures with (at pH = 7) and without pH control. In fact, the integral oxygen uptake rates for the two culture types are similar. The thermophilic populations present in the sludge do not show any activity during cultivation at pH values less than 6. The pH of the uncontrolled cultures normally increases to final values of approximately 8.8. These results indicate that the pH value does not have a significant influence on the response function for the process provided the pH value remains in the permissive window between > 6 and 8.8. It is therefore extremely important that the pH value of the raw sludge is known: when processes are operated without direct pH control, it will be necessary to assure that the feed of fresh sludge does not acidify the process by constrainting the feed rate (≈ in-

Table 5. Objectives and constraints for sewage sludge treatment processes.

Process objectives	Process constraints
Robust	continuous substrate supply
Rapid	feed volume and concentration variable
Low tech (minimal costs)	low cost
Reliable	expected load with toxic components
Flexible	variable substrate composition
Performance quantification	product quality must satisfy legal requirements

direct pH control). A pH value below 6 does not allow effective aerobic thermophilic treatment.

The aerobic thermophilic process involves spontaneous reaction under batch conditions. *B. stearothermophilus* strains are ubiquitous components of raw sludge at concentrations of approx. 10^4 colony-forming units (c.f.u.) ml^{-1}. They develop spontaneously when temperature and pH meet the physiological requirements described. The availability (preparation) of a constant active inoculum is of fundamental importance for spontaneous processes in order to achieve some reproducibility and to minimize unproductive lag times. Operating the process in the repetitive (fed) batch mode circumvents this problem. During experiments with this technique, major differences were observed with respect to the duration of the lag phase in the various cultivations. The lag phases are summarized in Table 6.

Not all of these cultures were oxygen limited, i.e., no plateau at the maximum Oxygen Uptake Rate (OUR_{max}) was attained, and they became carbon limited. In this case, the hydraulic retention time had a fundamental influence on the process efficiency. With frequent volume changes (every 3.5 h), part of the treated sludge is replaced with new feed close to the time where the maximum oxygen uptake rate occurs. Although the microorganisms have only a short lag phase on the new medium (1.8 h), such an operating procedure produces sludge that is not totally biodegraded; the organisms are never starved and, therefore, do not significantly tend to sporulate. When feeding is carried out less frequently (every 10.5 h), the process is more stable and shows a good reproducibility with respect to the oxygen uptake rate although the lag phase is longer (3.5 h).

This longer lag phase is related to partial sporulation of the process microflora during the carbon limited phase (Fig. 7). Partially responsible was also the temperature shock which, under these conditions, is greater than when volume changes occur every 3.5 or 7 h; heat exchange between treated and raw sludge was not possible in the experimental facility used (Bomio 1990). According to the trajectories of dipicolinic acid, a typical spore component, and the spore counts (determined on germination agar with valine as the spore-germination agent as described by Foerster 1983) sporulation occured immediately after the metabolically active phase in the carbon limited growth phase (in the example in Fig. 7: after 9 h cultivation time). Sporulation exerts a significant influence on the process performance.

The rates of reduction of the Chemical Oxygen Demand (COD) decreased as the volume change frequency decreased because of the longer lag phases. A fed-batch process with frequent volume changes is most effective with respect to the rate of destroying biodegradable organic matter. The cul-

Table 6. Dependence of the lag phases on the repetitive batch cycling conditions.

Time for volume changes [h]	3.5	7	10.5
Length of average lag phase [h]	1.8 ± 3.9	2.0 ± 1.3	3.5 ± 0.8

tivation time between volume changes during a fed batch process influences also microbiological heat production: frequent volume changes allow a greater heat production. This indicates strongly that a continuously operated process should be most efficient. However, continuous operation is said to be not acceptable; the argumentation is based on hygienic aspects and the fact that hydraulic short circuiting cannot be ruled out.

The heat produced by the metabolic activity of aerobic microorganisms (growth associated biogenic heat) is proportional to the oxygen uptake (Birou et al. 1987). The conversion factor is reported to be 440 kJ mol^{-1} oxygen independent of both, organisms and media. Now, the heat production rate was definitely lower in sludge cultivations with greater volume changes. This is a consequence of the more extended lag phase where no heat production occurs. The microbiological heat yield coefficient was essentially constant during cultivations at different scales and was found to be around 15 MJ kg^{-1} COD removal (see also McCarty 1965; Cummings & Jewell 1977; Breitenbücher 1983; Jewell & Kabrick 1980; Loll 1974; Kapp 1986).

Optimization
According to the definitions of objective functions, optimization experiments were carried out using a two-dimensional field of variables. This involved variation of the stirrer speed and the aeration rate because these two variables had a significant influence on process performance and on the chosen response function. Further, they are the only variables permitting the operator to vary with a considerable degree of freedom. The optimization according to the function that identifies the reduction of organic matter as a partial objective was repeated with different sludge batches in order to evaluate the practicability and reproducibility of the simplex method with a complex substrate such as sewage sludge (Fig. 8). A very sharp optimum was found in a laboratory scale reactor near a stirrer speed of 1500 min^{-1} with an air flow rate = 360 Nl h^{-1} (i.e. 1.5 vol/vol^{-1} · min^{-1}). The reproducibility of the simplex method was excellent and the first optimum was nicely reproduced in spite of beginning

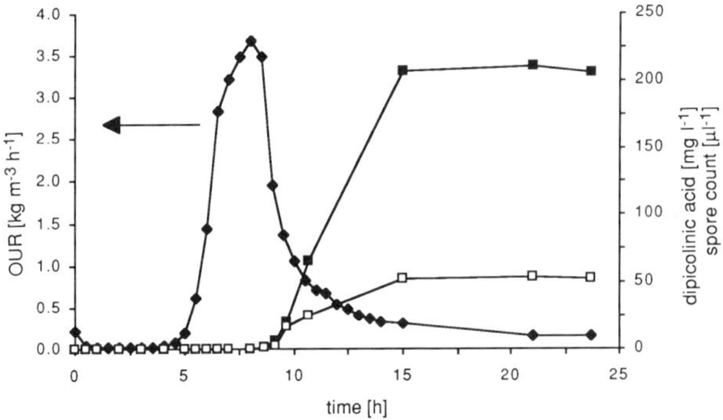

Fig. 7. Oxygen uptake rate (◆), spore count (□) and dipicolinic acid concentration (■) in sewage sludge during aerobic thermophilic sludge treatment in a lab scale bioreactor at 65° C, pH 7, stirrer speed 1500 min^{-1} and 1 vvm. In the carbon limited phase, sporulation of the thermophiles is observable by both spore count and by the detection of dipicolinic acid, a typical spore component.

the optimization with a different substrate and with considerably different starting simplex vertices (Bomio 1990). The optimum found seems to be independent of sludge composition.

The influence of the aeration rate was always positive; increased aeration resulted in increased response functions due to greater oxygen transfer rates. However, the mechanical power input (i.e. stirrer speed) did not have the same influence on process efficiency. Indeed, a too high stirrer speed sharply depressed the response function. *B. stearothermophilus* populations are not known to be shear sensitive, so the explanation of this event is not likely to be found in a damage of the single cells. The microbiological activity was found to be largely due to the metabolism of the particulate fraction. The degradation mechanisms correlate with intensive contact between microorganisms and the insoluble substrates, as demonstrated for cryptic growth (Hamer & Mason 1987). Most probably, excessively high mechanical power input (stirrer speed) does not allow an adequate contact between biomass and particulate substrates to be maintained or even destroys the aggregates and, hence, results in a drastic reduction of process efficiency. In general, the optimization results obtained with the simplex algorithm confirm the physiological knowledge about these populations

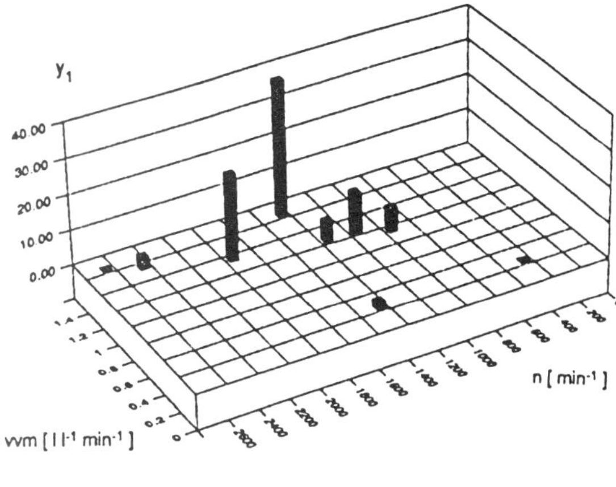

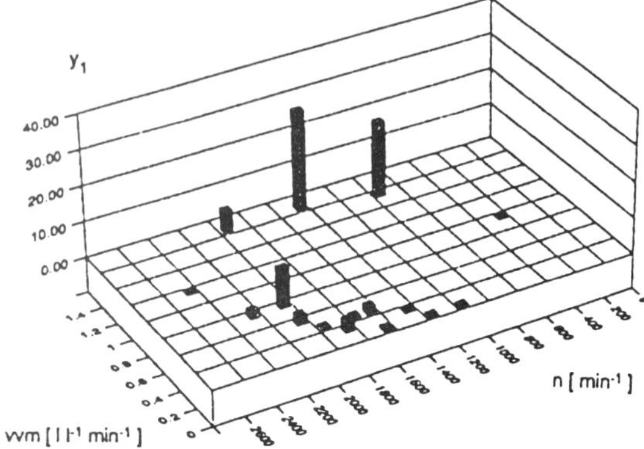

Fig. 8. Process optimization with the simplex algorithm; 2 different experimental series are shown. The maximum found is independent of both the raw sludge used and the starting vertices of the simplex. y_1 is an objective function consisting of: maximization of biogenic heat formation and COD removal, and minimization of biomass formation and process time; n is the stirrer speed in a 7 l COLOR; from Bomio 1990.

(i.e.: high potential for oxygen uptake and rapid growth).

Conclusions

The development of an efficient process for the treatment of waste sewage sludge involves definition and quantification of objective functions. In the case of a microbiological process in the wastewater treatment industry, this step requires synergistic interactions between scientific authorities, political and legislative authorities, and plant constructors. Reported results and unclear objectives attest that the synergistic effect is far from being fruitful.

The performance of aerobic thermophilic sludge treatment can be improved basing on augmented physiological knowledge about population dynamics in correlation with metabolism of different substrates in sewage sludge, i.e., studies of the biomass substrate relations must be intensified.

Aerobic thermophilic sludge treatment plants operating at 65° C take advantage of maximal thermophilic microbial respiratory activity and associated heat production. Temperatures less than 60° C are of little interest because of low hygienization efficiency. Temperatures greater than 70° C are too restricted with respect to microbial resources (the breadth of spectrum of organisms decreases rapidly with increasing temperature). The pH value is not a relevant factor for process efficiency provided it is not less than 6.8 because the permissive window is broad; pH regulation during the treatment process is not likely to be necessary, however, pH values as low as 5 . . . 6 often occur in untreated waste sludge. This must be either fed directly at a sufficiently low rate (in order not to acidify the process) or fed only after increasing pH by appropriate chemicals; the first solution is to be preferred. It is clear that the measurement of pH is an important process indicator.

Proteolytic activity is the dominating extracellular enzymatic activity found generally associated with the growth of thermophilic process microbes; it was always detected during an effective treatment. It obviously has an activity optimum at 80° C. The proteases present in the raw sludge were found to be completely inactive at this temperature. Therefore, it is possible to quantify the performance of a treatment unit by measuring the proteolytic activity at 80° C.

Extracellular proteases are the enzymes that are permanently active during normal operating conditions, but the presence of other easily degradable polymers in sludge, e.g. starch, rapidly induces the production of the respective degrading enzymes, in this example amylases.

Actively respiring thermophilic biomass can be quantified only with indirect methods such as total dehydrogenase activity. Such measurements allow the minimization of microbial growth during process development and optimization. Decreases in activity during batch and fed batch cultivations were found to be a consequence of oxygen or carbon limitation.

The heat inactivation of potentially pathogenic microorganisms is a pure temperature effect with no synergistic effects from the thermophilic populations. Neither antimicrobial agents nor lysozyme could be detected in sludge undergoing treatment. When potentially pathogenic microorganisms were added to ATS-processes their inactivation was slowed down compared with the death kinetics in pure aqueous medium; obviously, sludge components act protectively.

A great advantage of the aerobic thermophilic process over a purely thermal one (i.e. using microorganisms to generate heat) is the enhancement of sludge settling characteristics. This extra benefit is due to a decrease of small particles and an increase of particles larger than 30 μm during the process.

The repetitive (fed) batch cultivation technique is essential for the reproducibility of the process. Process harvest and feed patterns influence sludge biodegradation. Sporulation of thermophilic biomass is disadvantageous for the process because it results in long lag phases without significant microbial activity and must, therefore, be suppressed. Sporulation can be avoided by frequent volume changes, but this affects the level of stabilization achieved, i.e. high rates can so be maintained but the turn over is less than 100%. The yield coefficient – heat produced per COD removed – can be expected to be always close to 15 MJ kg^{-1}.

Continuing increases in sludge production in the future and increasing awareness of environmental problems might well stimulate research and development of technical scale sludge treatment processes using thermophilic microorganisms. Should this occur, the presented overview may give practical and scientific background knowledge for the design and construction of sludge treatment processes.

References

Baier U (1987) Zur Physiologie thermophiler *Bacilli*. PhD thesis ETH Zürich Nr 8423

Baier U, Sonnleitner B & Fiechter A (1986) Is physiological and genetical instability the limiting factor in biotechnological application of thermophilic pure cultures?: In: Alberghina L, Frontali L, Hamer G (Eds) Physiological and Genetic Modulation of Product Formation, Dechema Mongraph 105: 201–202

Bergquist PL, Love DR, Croft JR, Streiff MB, Daniel RM & Morgan WH (1987) Genetics and potential applications of thermophilic and extremely thermophilic microorganisms. Biotechnol Genetic Eng Rev 5: 199–24

Birou B, Marison IW & von Stockar U (1987) Calorimetric investigation of aerobic fermentations. Biotechnol Bioeng 30: 650–660

Bomio (1990) Bioprocess development for aerobic thermophilic sludge tretment. PhD thesis, ETH Zürich, No 9159

Bomio M, Sonnleitner B & Fiechter A (1989) Growth and biocatalytic activities of aerobic thermophilic populations in sewage sludge. Appl Microbiol Biotechnol 32: 356–362

Breitenbücher K (1983) Aerob-thermophile Stabilisierung von Abwasserschlämmen. PhD thesis, Hohenheim, Hohenheimer Arbeiten, Reihe Agrartechnik

Cummings RJ & Jewell WJ (1977) Thermophilic aerobic digestion of diary waste. In: Loehr RC (Ed) Food Fertilizer and Agricultural Residues (pp 637–657). Ann Arbor Science Press, Ann Arbor

Droste RL & Sanchez WA (1983) Microbial activity in aerobic sludge digestion. Water Res 17: 975–983

Feachem RG, Bradley DJ, Garelik H & Mara DD (1983) Sanitation and disease. Health aspects of wastewater management. World Bank Studies in Water Supply and Sanitation, Vol 3, 501 pp

Federal Department of the Interior (1981) Klärschlammverordnung vom 8. April (1981)/814.225.23, EDMZ Bern

Foerster HF (1983) Activation and germination characteristics observed in endospores of thermophilic strains of *Bacillus*. Arch Microbiol 134: 175–181

Gibson T & Gordon RE (1975) In: Buchanan RE, Gibbons NE (Eds) Bergey's Manual of Determinative Bacteriology, 8th edition (pp 539–540). Williams & Wilkins Comp, Baltimore

Hamer G & Mason CA (1987) Fundamental aspects of waste sewage sludge treatment: Microbial solids biodegradation in an aerobic thermophilic semi-continuous system. Bioproc Eng 2: 69–77

Hammel H-E (1983) Hygienische Untersuchungen über die Wirkung von Verfahren zur Kompostierung von entwässertem Klärschlamm und zur aerob-thermophilen Stabilisierung von Flüssigschlamm. PhD thesis, Justus-Liebig-Universität Giessen

Hamilton-Miller JMT (1977) Towards greater uniformity in sensitivity testing. J Antimicrobial Chemotherapy 3: 385–392

Hess P (1988) Untersuchungen zur Leistungsfähigkeit aerober

Suspensionskulturen in gerührten Bioreaktoren. PhD thesis, ETH Zürich, Nr 8572

Holm-Hansen O (1973) The use of ATP determinations in ecological studies. Bull Ecol Res Comm (Stockholm) 17: 215–222

Jewell WJ & Kabrick RM (1980) Autoheated aerobic thermophilic digestion with aeration. J WPCF 52: 512–523

Kapp H (1986) Aerobe thermophile Klärschlammbehandlung vor der anaeroben Stabilisierung. Korrespondenz Abwasser 33: 1038–1042

King JD & White DC (1977) Muramic acid as a measure of microbial biomass in estuarine and marine samples. Appl Environ Microbiol 33: 777–783

Kresken M & Wiedmann B (1984) MIC reading with the biology laboratory computer. In: Habermehl K-O (Ed) Rapid Methods and Automation in Microbiology and Immunology (pp 490–496). Springer Berlin

Lentner C (1981) Geigy Scientific Tables. Vol 1, units of measurement, body fluids, composition of the body, nutrition, 230

Loll U (1974) Stabilisierung hochkonzentrierter organischer Abwässer und Abwasserschlämme durch aerob-thermophile Abbauprozesse. PhD thesis Technische Hochschule Darmstadt

Lopez JM, Koopman B & Bitton G (1986) INT-dehydrogenase test for activated sludge process control. Biotechnol Bioeng 28: 1080–1085

Mason CA, Hamer G & Bryers JD (1986) The death and lysis of microorganisms in environmental processes. FEMS Microbiol Rev 39: 373–401

McCarty PL (1965) Thermodynamics of biological synthesis and growth. Adv Water Poll Res 2(2): 169–187

Nebiker H (1981) Flüssigrotte als Grundlage stabiler Schlammhygienisierung durch natürlichen Antagonismus. Chemische Rundschau 20: 2–7

Schweizer HU (1988) Rechtliche Grundlagen der Klärschlammentsorgung. 20. VSA-Fortbildungskurs, Klärschlamm: Behandlung-Verwertung-Entsorgung, Engelberg, CH

Sonnleitner B & Fiechter A (1983a) Bacterial diversity in thermophilic aerobic sewage sludge: I. Active biomass and its fluctuations. Eur J Appl Microbiol Biotechnol 18: 47–51

(1983b) Thermophilic microflora in aerated sewage sludge. In: Processing and use of sewage sludge, Conference at Brighton 27–30, 1983 (pp 235–236)

(1983c) Bacterial diversity in thermophilic aerobic sewage sludge: II. Types of organisms and their capacities. Eur J Appl Microbiol Biotechnol 18: 174–180

Sundaram TK (1988) Thermostable enzymes for biotechnology. J Chem Tech Biotechnol 42: 308–311

Thomanetz E (1982) Untersuchungen zur Charakterisierung und quantitativen Erfassung der Biomasse von belebten Schlämmen. Stuttgarter Berichte zur Siedlungswasserwirtschaft, Vol 74, 183 pp

Viel M, Sayag D, Peyre A & André L (1987) Optimization of in-vessel co-composting trough heat recovery. Biological Wastes 20: 167–185

Wassen H (1975) Hygienische Untersuchungen über die Verwendbarkeit der Umwälzbelüftung (System FUCHS) zur Aufbereitung von flüssigen Abfällen aus dem kommunalen und landwirtschaftlichen Bereich. PhD thesis Justus Liebig-Universität, Giessen

Walker NE & Campbell LL (1966) Purification and properties of a lytic enzyme from induced cultures of *Bacillus stearothermophilus*. (TP-1). Bacteriol Proc Am Soc Microbiol 66: 126–126

White DC, Bobbie RJ, Herron JS, King JD & Morrison SJ (1979) Biochemical measurements of microbial biomass and activity from environmental samples. In: Costerton JW, Colwell RR (Eds) Native Aquatic Bacteria, Enumeration, Activity, and Ecology (pp 69–81). American Society for Testing and Materials, Philadelphia

Biodegradation **1**: 147–161, 1990.

Enzymology of cellulose degradation

Thomas M. Wood & Vicenta Garcia-Campayo
Rowett Research Institute, Greenburn Road, Bucksburn, Aberdeen AB2 9SB, UK

Key words: fungal cellulase, bacterial cellulase, mode of action, structure/activity relationships, synergism

Abstract

In the last few years there has been a considerable improvement in the understanding of the mechanisms involved in the microbial degradation of cellulose, but there are still many uncertainties. As presently understood, it would appear that different mechanisms may operate in the various types of microorganism. Thus degradation of crystalline cellulose is effected by anaerobic bacteria by large Ca-dependent and thiol-dependent multicomponent endoglucanase-containing complexes (cellulosomes) located on concerted action of endo- and exo-glucanases which act some distance from the cell which renders cellulose soluble. All of the endo- and exo-glucanases possess a bifunctional domain structure: one contains the catalytic site, the other is involved in binding the enzyme to crystalline cellulose.

Introduction

It would appear that a number of different mechanisms operate in the microbial solubilization of crystalline cellulose. Cellulolysis by soft rot and white rot aerobic fungi (Eriksson & Wood 1985) and some aerobic bacteria (Coughlan 1990) involves the synergistic action of enzymes, loosely defined as exoglucanases (normally cellobiohydrolases i.e. 1,4-β-D-glucan cellobiohydrolase), endoglucanases (endo-1,4-β-D-glucan-4-glucanohydrolase) and β-glucosidases (Eriksson & Wood 1985). Brown rot fungi, on the other hand, produce endoglucanases but no exoglucanases and may have a different mechanism, perhaps involving H_2O_2 (Koenings 1975). Some anaerobic bacteria (Lamed & Bayer 1988) and possibly anaerobic fungi (Wood et al. 1988) use a multicomponent enzyme complex which contains endoglucanases, but the exact composition of the complex remains to be described in each case.

Unfortunately, space does not permit a discussion of all these mechanisms. Of necessity, the review will focus mainly on those mechanisms and microorganisms which have been the subject of most research activity, namely, the aerobic fungi and the anaerobic bacteria. Moreover, discussion of fungal cellulases is restricted only to enzymes classified as endoglucanase and exoglucanase (cellobiohydrolase): enzymes such as glucohydrolase, cellobiose oxidase and cellobiose dehydrogenase, which are found in some culture filtrates as minor constituents (Eriksson & Wood 1985), are not discussed further. Thus this review is confined mainly to a discussion of those enzymes and enzymatic processes by which crystalline hydrogen bond-ordered cellulose is rendered soluble.

Much progress has been made in understanding the mechanism by which fungi and bacteria degrade hydrogen bond-ordered cellulose, but there are still many uncertainties. Some of the most recent interesting insights into the mechanism have been obtained from studies of the structure/function relationships provided by structural analysis of the enzymes, by studies on cloned genes and on enzymes resulting from expression of these cloned genes.

Fungal cellulases

Composition of the cellulase system

The distinguishing feature of the cellulase that can solubilize crystalline cellulose is that it contains a cellobiohydrolase. This is in addition to the randomly-acting endoglucanases and β-glucosidases/cellobiases found in all fungal culture filtrates. Only a few fungi synthesise and release into the culture medium appreciable amounts of the cellobiohydrolase enzyme. Notable in this regard are the fungi, *Trichoderma reesei, Trichoderma viride, Fusarium solani* and *Penicillium funiculosum/pinophilum* (Eriksson & Wood 1985).

All fungal cellulases studied so far have been shown to contain a multiplicity of enzyme components (Wood 1990; Coughlan 1985). The actual number of components depends on the source of the fungus and the manner in which it has been cultured. *Trichoderma viride* and *Trichoderma reesei* cellulases have been most extensively studied (Eriksson & Wood 1985; Wood 1990; Coughlan 1985). They have been shown to contain four to eight endoglucanases, two cellobiohydrolases and one to two β-glucosidases (Coughlan 1985; Wood 1990). *Penicillium funiculosum/pinophilum* cellulase contains two cellobiohydrolases (Wood et al. 1980; Wood & McCrae 1986a), five to eight endoglucanases (Bhat et al. 1989) and two β-glucosidases (Wood et al. 1980). Other cellulases are equally heterogeneous (Streamer et al. 1975; Wood 1990). It seems that only some of these components are genetically determined; others are artefacts resulting from differential glycosylation of a common polypeptide chain (Wood & McCrae 1972; Gum & Brown 1977), from partial proteolysis (Eriksson & Pettersson 1982), from aggregation of the enzymes with each other or with part of the fungal cell wall (Sprey & Lambert 1983), or from manipulation of the enzymes during purification (Enari & Niku Paavola 1987). These artefacts make elucidation of the mechanism of action extremely difficult: consequently there is considerable discussion on the substrate specificity of the enzymes, on the mode of action of the individual enzymes, particularly the cellobiohydrolases, and on the nature of the co-operation between the various enzymes. Currently, there is some agreement that the *extensive* conversion of crystalline cellulose to glucose can be discussed in terms of the co-operative action of two immunologically unrelated cellobiohydrolases (so-called CBH I and CBH II), one or more randomly-acting endoglucanases and at least one β-glucosidase (Wood 1990).

The problem of classification of the enzymes

Table 1 summarizes views that have been widely held on the substrate specificities of the enzymes which are found in the cellulases that can degrade crystalline cellulose. However, as the properties of the isolated enzymes have been studied more closely, it is becoming increasingly clear that it is difficult to classify them strictly as endoglucanases and cellobiohydrolases. For example, some purified cellobiohydrolases are reported to attack barley β-glucans (Henrissat et al. 1985) and even CM-cellulose (Wood 1990), which are substrates long been held to be degradable only by enzymes classified as randomly-acting endoglucanases. On the other hand, some enzymes classified as endoglucanases are reported to be able to hydrolyse crystalline cellulose (Beldman et al. 1985; Enari & Niku-Paavola 1987), which of course is presumed to be a property of the

Table 1. Action of cellulase components on different substrates.

Enzyme	Crystalline cellulose	Amorphous 'swollen' cellulose	CM-Cellulose	Cello-oligosaccharides	Cellobiose
Cellobiohydrolase	Slow	Very active	Nil	Active	Nil
Endoglucanase	Nil	Very active	Very active	Active	Nil
β-Glucosidase	Nil	Nil	Nil	Active	Active

cellobiohydrolase. But the confusion does not end there: yet another endoglucanase is reported to have no action on amorphous cellulose prepared by milling cellulose powder in ethanol (Niku-Paavola et al. 1985). There are other apparent anomalies too frequent to report here.

Clearly, the substrate specificities of the various enzymes continues to be a contentious issue. However, the explanation may be quite simple, viz, either the enzymes have overlapping substrate specificities or some of the enzymes which have been reported to be pure obviously are not. The purity of the cellobiohydrolases in particular has been questioned and, as a consequence, the mode of action of these enzymes has been the subject of great debate.

Are the cellobiohydrolases endo- or exo-acting?

It has been held for many years that cellobiohydrolases (CBHs) are exoglucanases that remove cellobiose consecutively from the non-reducing end of the cellulose chain. Typically, crystalline and amorphous cellulose have been reported to be degraded to cellobiose, the rate depending on the degree of polymerisation, and the crystallinity of the cellulose (Wood & McCrae 1979; Wood 1990). Recently, however, there have been several reports which have indicated that the cellobiohydrolases may not attack exclusively the penultimate glycosidic link at the non-reducing end of the cellulose chain. Unfortunately, the case is weakened by the apparent lack of agreement as to whether it is CBH I or CBH II or both that possess this property. Thus, a CBH I from *T. reesei* attacked barley β-glucan in a random manner typical of an endoglucanase (Henrissat et al. 1985), but CBH I from *P. pinophilum* effected only a slow change in the degree of polymerisation of the β-glucan (Wood et al. 1989), as would be expected from an exo-acting enzyme. On the other hand, another CBH I from *T. reesei* hydrolysed chromophoric cello-oligosaccharides in a manner not typical of an exoglucanase (Claeyssens et al. 1989).

Support that the CBH I enzyme may not act from the end of the chain has been obtained by electron microscopy. Thus, Chanzy et al. (1984) noted that CBH I from *T. reesei*, labelled with colloidal gold, was found to be attached to microcrystals of the alga, *Valonia microphysa*, along the length of the microfibril. White & Brown (1981), in a similar study, but using cellulose from the bacterium *Acetobacter xylinum*, made the same observation.

Conclusions as to the mode of action of CBH II are equally diverse. Thus, electron microscopic evidence showing that CBH II from *T. reesei* attacked *Valonia* cellulose microcrystals only from the non-reducing end supports the claim that it is a true exoglucanase (Chanzy et al. 1985). Using biochemical studies, a similar mode of action was deduced for a CBH II from *P. pinophilum* purified by affinity chromatography (Wood et al. 1989). By contrast, Enari & Niku Paavola (1987) and Kyriacou et al. (1987), also using biochemical studies, conclude that CBH II from *T. reesei* are endo-acting, albeit 'less randomly-acting' than a typical endoglucanase.

How can these conflicting results be rationalised? Clearly, enzymes from different sources may indeed have different substrate specificities; and some of the variations at least may therefore be quite easily explained. However, when the same enzyme from the same source would appear to have completely different properties, another reason must be sought. One possibility is that the apparent differences may be a consequence of the existence of aggregates or enzyme-enzyme complexes between cellobiohydrolases and endoglucanases that are extremely difficult to break into their constituent parts. Such complexes have been shown to exist in a cellulase from *T. reesei*. In this case, electrophoretically homogeneous complexes between endoglucanase, xylanase and β-glucosidase were found to be heterogeneous after treatment with a urea/octyl glucoside dissociation reagent (Sprey & Lambert 1983). Similar complexes have been found to exist between cellobiohydrolases and endoglucanases in electrophoretically homogeneous enzyme preparations isolated from cultures of *P. pinophilum* and *T. reesei* (Wood et al. 1989). These complexes, which were homogeneous after extensive purification involving gel filtration, chromatofocusing and affinity chromatography on a column of cellulose, were found to be heterogeneous after affinity chromatography on a column that had been prepared by coupling *p*-aminobenzyl-1-thio-β-D-cellobioside to Affigel 10 (van Tilbeurgh et al. 1984). A CBH II from *P. pinophilum* purified in this way had the properties of a typical exoglucanase: the enzyme could effect only a slow decrease in the viscosity (a parameter related to chain length) of a solution of barley β-glucan in contrast to the rapid decrease shown by purified endoglucanases from the same fungus: CBH I and II from *T. reesei* prepared on the cellobioside column were similar in this respect (Tomme et al. 1988a).

Whether or not enzyme-enzyme complexes can explain the differing opinions on the substrate specificity and mode of action, it is clear that at least some of the confusion regarding the properties of the cellobiohydrolase enzymes is caused by the difficulty in obtaining single enzyme species. Expression of the cellulase genes in a heterologous host makes it possible to produce each enzyme free of contaminating glycosidases. It may therefore be significant that a CBH I gene from *T. reesei*, expressed in yeast, showed no capacity to hydrolyse barley β-glucan (Knowles et al. 1988b), which is readily degraded by endoglucanases (Wood et al. 1989), while a recombinant CBH II did. The implication is that CBH I from *T. reesei* may indeed be an exocellobiohydrolase and CBH II may have some endo-type action. The fact that the recombinant CBH II had no apparent activity to CM-cellulose, which is typical of an endoglucanase (Table 1), however confuses the issue. Perhaps this observation will enforce a more critical evaluation of the use of non-cellulosic substrates (barley β-glucan) and unnatural substrates (CM-cellulose) in studies of the mode of action.

Chromophoric cello-oligosaccharides as substrates

To some extent, the difficulty in reaching a consensus on the mode of action of the cellulase components is a direct consequence of having to use substrates which are poorly characterized and assay methods which lack sensitivity. In an attempt to overcome these problems, the use of chromophoric glycosides from cello-oligosaccharides and lactose substrates in conjuction with fluorescence and HPLC methods has been pioneered by van Tilbeurgh et al. (1982). This elegant procedure, in demonstrating the preferred site of hydrolysis, has provided interesting insights into the mode of action of CBH I and II of *T. reesei* and *P. pinophilum* (Claeyssens et al. 1989). Thus the results have been interpreted to indicate that the site of attack of all four cellobiohydrolase enzymes is not restricted to the penultimate glycosidic linkage at the non-reducing end, suggesting that there is a degree of randomness in the attack. However, as the CBH I and II of *P. pinophilum* produced cellobiose almost exclusively when cellulose swollen in H_3PO_4 was the substrate, and effected only a slow fall in the degree of polymerisation, one clearly must be cautious in drawing conclusions on the mechanism of action from only one substrate (Wood et al. 1980; Wood & McCrae 1986a, b).

Despite these reservations, a range of chromophoric glycosides has been extremely useful in classifying the cellulase components of *T. reesei*. Thus, 4-methylumbelliferyl cellobioside and the corresponding lactoside are hydrolysed by CBH I or an endoglucanase designated EG I (Claeyssens & Tomme 1989), but not by CBH II or EG III (Saloheimo et al. 1988). The structure/activity relationships discussed below show the significance of this classification.

Thus the controversy as to the mode of action and the substrate specificity of the components of the cellulase system of fungi continues.

Synergism

When the observed action of two or more enzymes acting together in solution is greater than the sum of the individual actions, it is concluded that the enzymes act synergistically. There are two types of synergistic action involved in the process by which crystalline cellulose is rendered soluble, i.e. co-operation between endoglucanase and cellobiohydrolase (so-called, endo-exo synergism) (Wood & McCrae 1972; 1979) and co-operation between two cellobiohydrolases (so-called, exo-exo synergism)

(Fagerstam & Pettersson 1980). However, despite intense research activity the molecular basis for synergistic action is not well understood. It is possible that the lack of agreement is a direct consequence of the wide diversity of opinion regarding the individual roles of the 'purified' enzymes. However, the choice of substrate as an example of crystalline cellulose has also been responsible for some confusion. This was elegantly demonstrated by Henrissat et al. (1985) who showed that the degree of synergistic activity observed between cellobiohydrolase and endoglucanase varies with substrate used as well as with the ratio of cellobiohydrolase to endoglucanase. It appears, however, that synergism is most marked when crystalline cellulose is the substrate, that it is low or non-existent with amorphous-highly hydrate cellulose, and that it is absent with soluble cellulose derivatives (Wood & McCrae 1979).

The original model for synergistic activity between enzyme components envisaged an enzyme (so-called C_1) whose sole function was to cause some relaxation in the intramolecular hydrogen bonding as a preliminary to action by the hydrolytic enzymes (Reese et al. 1950). No one now believes that such an enzyme exists, but there is no doubt that the disaggregation and subsequent hydration of the closely packed cellulose chains in the cellulose crystallite is an essential prerequisite of cleavage of the glycosidic bond by cellulase enzymes. The discovery that cellobiohydrolase and endoglucanase consist of two domains, one binding and one hydrolytic (Tomme et al. 1988b) suggests that the swelling and hydrolytic function may reside in one enzyme. Knowles et al. (1988) envisage the binding domain to 'unzip' the individual chains as a preliminary to hydrolysis of the cellulose by the hydrolytic domain. This suggestion though relates only to action by one enzyme.

Attempts to explain the concerted action of endoglucanase and cellobiohydrolase have discussed the mechanism in terms of sequential action, where a randomly acting endoglucanase initiates the attack in the amorphous areas of the cellulose to create non-reducing ends for the endwise-acting cellobiohydrolase (Wood & McCrae 1972). However, this model is granted only qualified acceptance and it is generally regarded as an oversimplification. It does not account, for example, either for the fact that there is little or no synergism observed between some endoglucanases and cellobiohydrolases (Wood 1975) or that synergism exists between two cellobiohydrolases (Fagerstam & Pettersson 1980).

There is no doubt that adsorption of the enzyme on the cellulose is essential for solubilisation (Coughlan 1985; Klyosov 1988). Klyosov and his colleagues (Klyosov 1988) conclude that only those endoglucanases that have a strong affinity for crystalline cellulose can act synergistically with the cellobiohydrolase. Ryu et al. (1984) are of the opinion that exo-endo synergism can be described in terms of competitive adsorption of the two types of enzyme, optimum co-operation being manifested when the enzymes were present in the ratio in which they were present in the culture filtrate. Woodward et al. (1988a, 1988b), on the other hand, showed that the concentration of the mixture of cellobiohydrolase and endoglucanase from *T. reesei* was more important than the ratio of the enzymes. As the concentration of the enzyme mixture decreased, the degree of synergism increased to a maximum and then decreased with further increases in enzyme concentration.

Fagerstam & Pettersson (1980) were the first to demonstrate that purified CBH I and CBH II from *T. reesei* co-operated to effect hydrolysis of crystalline cellulose. This unexpected finding was confirmed by Henrissat et al. (1985) and Kyriacou et al. (1987) working on the same cellulase and by Wood & McCrae (1986b) using CBH I and II from *P. pinophilum*. Henrissat et al. (1985) envisaged that competitive adsorption or the formation of a binary complex between CBH I and CBH II might result in the enhancement of the turnover of the enzymes. There is now some evidence that such complexes do in fact exist (Tomme et al. 1990).

Synergism between two cellobiohydrolases acting at the ends of the substrate is difficult to explain. Wood & McCrae (1986b) have postulated that the mechanism can be discussed in terms of the stereochemistry of the cellulose chains, based on the fact that there are likely to be two different naturally-occurring configurations of non-reducing end group in the cellulose crystallite. In essence, they envisage that CBH I and CBH II may differ in their substrate stereospecificities, and that the apparent cooperation may be discussed in terms of CBH I attacking only one of the two stereospecifically-different non-reducing end groups, while CBH II attacks the other. Thus, synergistic action would be observed if the sequential removal of cellobiose from one type of non-reducing end by CBH I exposed, on a neighbouring chain, a non-reducing chain of different configuration which would be a substrate for CBH II, and vice versa.

Clearly, there is a great deal of debate and uncertainty regarding the mechanism of synergistic action between the various enzymes and rationalisation is difficult. However, it is abundantly clear that there can be no agreement on the matter until there is a consensus of opinion regarding the substrate specificity and mode of action of the cellobiohydrolases. As purification techniques improve, so views on the properties of the individual enzymes may be modified. Of interest in this regard has been the use of affinity chromatography on a column of *p*-aminobenzyl-1-thio-β-cellobioside, which is a technique developed by Van Tilbeurgh et al. 1984). Thus, as mentioned above, Wood et al. (1989) have shown that preparations of *P. pinophilum* CBH II, isolated by conventional separation techniques including affinity chromatography on a column of cellulose, and shown to be electrophoretically homogeneous, were in fact contaminated by trace amounts of endoglucanase. When the contaminating endoglucanase was removed on the affinity column, no synergism was observed between CBH I and II nor between a mixture of CBH I and II and either of the endoglucanases, when crystalline cellulose in the form of the cotton fibre was used as substrate. Synergistic action was only apparent when CBH I and II and a specific endoglucanase were present in admixture (Fig. 1). The optimum ratio of the cellobiohydrolase components was 1 : 1 and, significantly, the addition of a trace of endoglucanase was needed for extensive degradation of the substrate. Thus, it appears that when the enzymes are highly purified, three enzymes, and not two as previously believed, are required for a reasonable rate of hydrolysis of crystalline cellulose in the form of cotton fibre.

The implication of these new results is that many of the contradictory statements in the literature may be the result of incomplete resolution of enzyme complexes. The authors report (Wood et al. 1989; Wood 1989) that an electrophoretically homogeneous preparation of *T. reesei* CBH II, prepared in their laboratory, could be further purified in the same way. Cellobiohydrolase preparations from *T. reesei* purified in different ways have been shown to be able to degrade cotton fibre without the need for the addition of endoglucanase (Enari & Niku Paavola 1987): perhaps, in this case, one or other of the enzymes were complexed with small amounts of contaminating endoglucanase which acted synergistically with the cellobiohydrolase.

Bacterial cellulases

In comparison with the cellulases of the fungi, very little is known about the mechanisms by which bacteria degrade cellulose. To a large extent this is a consequence of the fact that many bacteria, unlike fungi, degrade the cellulosic fibre by erosion of the surface and use cell-bound enzymes. It appears possible, however, that 'fungal-like' prokaryotes such as the *Actinomycetes* and the *Corynebacteria* might degrade cellulose using a mechanism similar to that operating in the fungi (Beguin 1990). It has been suggested that this can be rationalized to indicate that cell-bound enzymes will be more efficient in some situations (Yablonsky et al. 1988). These include situations where the microorganism will be exposed to predatory microorganisms, or where it is operating in an aquatic environment or in ecosystems such as that in the rumen. Cell-free enzymes, on the other hand, would be more effective in aiding the spread of mycelia through the plant cell wall by predigestion by the extracellular enzymes.

Cell-bound enzymes are more difficult to study. Some of the most efficient cellulolytic bacteria, such as *Sporocytophaga myxococcoides*, release practically no extracellular enzyme. Most bacteria

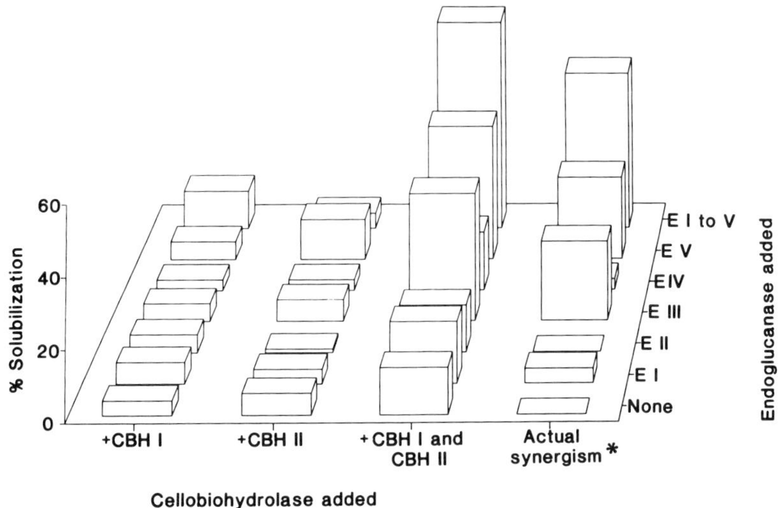

Fig. 1. Synergism between cellobiohydrolases and endoglucanases in solubilizing cellulose (cotton). Actual synergism means the solubilization effected by a mixture of CBH I and CBH II minus the solubilization recorded by a mixture of the two cellobiohydrolases. From: Wood 1989

that do secrete cellulases, in fact release only a variety of endoglucanases which show little activity to crystalline cellulose. Exceptionally, cell-free enzymes with activity towards crystalline cellulose are found in cultures of the anaerobes: *Clostridium thermocellum* (Lamed et al. 1985; Johnson et al. 1982), *Acetivibrio cellulolyticus* (McKenzie et al. 1987) and *Clostridium stercorarium* (Creuzet et al. 1983), and the aerobe *Microbispora bispora* (Yablonski et al. 1988).

The activity of cellulases from *C. thermocellum* (Johnson et al. 1982) and *A. cellulolyticus* (McKenzie et al. 1987) towards crystalline cellulose appears to depend on the presence of Ca^{2+} and thiol reducing agents (Johnson et al. 1982). In the case of *C. thermocellum*, activity towards crystalline cellulose is found in a tightly associated multimolecular complex containing endoglucanases but, as far as is known, no cellobiohydrolase. Cellobiohydrolases (or at least exoglucanases) have been identified in culture filtrates of *Cellulomonas fimi* (Miller et al. 1988), *Clostridium stercorarium* (Creuzet et al. 1983), *Ruminococcus flavefaciens* (Gardner et al. 1987), *Microbispora bispora* (Yablonsky et al. 1988), *Streptomyces flavogriseus* (McKenzie et al. 1984), on the basis of their capacity to hydrolyse nitrophenyl- or 4-methylumbelliferyl-cellobioside, the corresponding lactosides, or amorphous cellulose, but only two (*C. stercorarium; M. bispora*) have been characterised as 'fungal-like' cellobiohydrolases, which release cellobiose virtually exclusively from crystalline cellulose and which act synergistically with the endoglucanases in solubilizing crystalline cellulose.

Only a few bacterial extracellular cellulases have been studied in any detail: of those, *C. thermocellum* cellulase has received most attention (Lamed & Bayer 1988). The multicomponent cellulase complex of *C. thermocellum* is a very stable structure, comprising 14 to 18 polypeptides (Lamed et al. 1983a, b). The attachment of the complex, which has been termed the cellulosome (Lamed et al. 1983b), may be mediated by a non-cellulolytic binding factor (S_L) of molecular mass 250 kDa (Wu et al. 1988). Polycellulosomal protuberances cover the surface of the bacterial cell at periodic intervals (Bayer & Lamed 1986; Lamed & Bayer 1988). The cellulosomes dissociate from the cell after a time and are to be found in clusters covering the whole of the residual cellulose (Mayer 1988; Mayer et al. 1987). A number of other aerobic, anaerobic, mesophilic and thermophilic cellulolytic bacteria have been shown to have these cellulosomes (Coughlan & Ljungdahl 1988; Beguin 1990): however, no polycellulosomal protuberances are found on non-cellulolytic bacteria. Each cellulosome is characterized by the presence of a polypeptide of molecular mass 200 to 220 kDa (Coughlan & Ljungdahl 1988; Ljungdahl 1989), which may be involved in binding the cellulosome to the cellulose. Thus, bacteria which utilise a multicomponent cellulase for hydrolysis of crystalline cellulose have similar structural features and enzymic properties.

The cellulosome of *C. thermocellum* can be fragmented into a number of subunits using SDS under different conditions of temperature and concentration (Lamed & Bayer 1988; Bayer et al. 1985; Lamed et al. 1985). One subunit (S1) is of special interest (Lamed et al. 1983a). It is highly antigenic, it contains about 40% carbohydrate, it is non-cellulolytic and it is relatively easily removed from the rest of the cellulosome by low concentrations of sodium dodecylsulphate (SDS) (Lamed & Bayer 1988). A cellulose-binding role has been considered for S1. With higher concentrations of SDS, the intact cellulosome dissociates readily into a number of subunits, some of which show endoglucanase activity after renaturation, but activity to crystalline cellulose is lost completely even when the dissociating reagents are removed and a multicomponent cellulase complex is reformed (Lamed & Bayer 1988). Using fairly drastic conditions, Wu et al. (1988) were able to effect the dissociation of the cellulosome into a number of components, two of which had M_r value 82 kDa (designated S_s) and 250 kDa (S_L), respectively. Only S_s showed CM-cellulase activity, but neither could degrade crystalline cellulose in the form of Avicel when acting alone. However, when S_s and S_L were recombined some of the activity was recovered. Neither Ljungdahl (1989) or Lamed & Bayer (1988) have isolated a subunit of this molecular weight; consequently it is not clear if this S_L relates to subunit S1.

A yellow affinity substance (YAS), which has a high affinity both for the cellulose and the cellulosome, also appears to have a role to play in binding (Ljungdahl et al. 1983). Ljungdahl et al. (1988) speculate that YAS may be a factor used by the bacterial cell to recognize cellulose. The structure of the YAS is not known precisely, but there is some evidence that it may be a carotinoid (Ljungdahl 1989).

It is obvious that the mechanism by which the cellulosome hydrolyses cellulose is not well understood, but it would appear to be quite different to that operating in the aerobic fungi. On the basis of evidence obtained in the electron microscope, Mayer et al. (1987) have proposed a model (Fig. 2) in which the subunits of the cellulosome attack the cellulose chains simultaneously at every eight glycosidic linkage to release cello-oligosaccharides which are four cellobiose (C_4) units in length. This model was deduced when it was observed that the average distance between the polypeptides was 4 nm, which corresponds in length to a cello-oligosaccharide of four cellobiose units. Subsequently, the C_4 units may be degraded to cellotetraose or cellobiose by rows of subunits which are smaller in size. The proposers do not make it clear how the large variety of endoglucanases resulting from expression of the multitude of genes found in C. thermocellum fit into this model.

Structure/activity relationships in fungal and bacterial cellulases

All cellulolytic microorganisms of fungal or bacterial origin which have been studied so far contain multiple genes. Remarkably, fifteen different endoglucanase genes, two xylanase genes and two β-glucosidase genes of *C. thermocellum* have been cloned in *E. coli* (Hazlewood et al. 1988; Grabnitz & Staudenbauer 1988) and several have been sequenced [celA, celB, celC, celD, celF, xynZ and bglB – (Beguin 1990)]. The products of these have been purified and one (endoglucanase D) has been crystallised (Joliff et al. 1986). Other bacteria contain a similar multiplicity in the genes. Thus, 10 have been cloned from *Ruminococcus albus* 8 (Howard & White 1988). Other examples of multiple genes in bacteria have been found in *M. bispora* (Yablonsky et al. 1988, 1989), *Erwinia chrysanthemi* (Chippaux 1988) and *C. fimi* (Miller et al. 1988). Of the cellulolytic fungi, the molecular genetics of

T. reesei have been studied in most detail and two cellobiohydrolases [CBH I (Shoemaker et al. 1983; Teeri et al. 1983), CBH II (Teeri et al. 1987b; Chen et al. 1987)] two endoglucanases [EG I (Penttila et al. 1986; van Arsdell et al. 1987) and EG III (Saloheimo et al. 1988)] have now been cloned, sequenced and the primary structure deduced.

A comparison of primary protein sequences derived from the nucleotide sequences has shown that cellulase components from different sources have a common design in both fungi and bacteria. Thus, all cellulase components from these widely different sources have a common design in both fungi and bacteria. Thus, all cellulase components appear to be composed of two separate functional domains, which are conserved to different degrees and integrated into the proteins in different orders. Each enzyme appears to consist of a non-conserved catalytic core protein which is linked by a flexible hinge region, usually rich in proline, threonine and serine, to a highly conserved tail region which is situated at either the C- or N-terminal end of the molecule.

The domain structure has now been characterized to different degrees in the fungi *T. reesei* (Knowles et al. 1988b; Teeri et al. 1987b) and *Sporotrichum pulverulentum* (Johanson et al. 1989), and there is some evidence for it in *P. pinophilum* (Claeyssens & Tomme 1989). Among the bacteria, it has been demonstrated in *C. fimi* (Warren et al. 1986; Gilkes et al. 1988), *M. bispora* (Yablonsky et al. 1988), *Bacteroides succinogenes* (McGavin & Forsberg 1989), *C. thermocellum* (Hall et al. 1988; Faure et al. 1989) and *Thermononospora fusca* (Changas et al. 1988).

Studies on the domains of *T. reesei* cellulase components have been particularly thorough (Shoemaker et al. 1983; Teeri et al. 1983; Pentilla et al. 1986; Chen et al. 1987; van Arsdell et al. 1987; Teeri et al. 1987a; Teeri et al. 1987b; Saloheimo et al. 1988). Thus, it has been shown that CBH I, CBH II, EG I and EG III contain highly conserved regions (designated A, B in Fig. 3) which are either at the C-terminal (CBH I, EG I) or the N-terminal (CBH II, EG III) end of the enzymes (Fig. 3). Block A, which comprises approx. 30 amino acid residues is the better conserved in all four components (70% homology). This block is rich in glycine and cysteine and is stabilized by 2 or 3 disulphide bridges (Bhikhabai & Pettersson 1984): it exists as a small domain which is separate from the core or catalytic domain. The domain structure of CBH I (Schmuck et al. 1986; Abuja et al. 1988a) and CBH II (Abuja et al. 1988b) was shown to consist of a large ellipsoid head and a long tail reminiscent of a tadpole (Fig. 4). The small domain (block A) is joined to the rest of the protein by region B (Fig. 4). This region, which is rich in proline, serine and threonine (approx 40%) is heavily *O*-glycosylated (Fagerstam et al. 1984; Bhikhabhai & Pettersson 1984; van Tilbeurgh et al. 1986; Tomme et al. 1988b), which may protect against proteolytic attack (Claeyssens & Tomme 1989). Homology in region B is 50–60%. This region is assumed to function as a flexible hinge linking the small terminal domain to the larger domain (Knowles et al. 1988b). A recombinant endoglucanase (ngCenA) from *C. fimi* expressed in *E. coli* has a gross structural and functional organization similar to that of CBH I and CBH II of *T. reesei*, but the model incorporates a constrained angle of 135° between the long axes of the core and tail regions (Pilz et al. 1990). The binding domain of the recombinant enzyme could be excised precisely by an extracellular *C. fimi* protease (Gilkes et al. 1988).

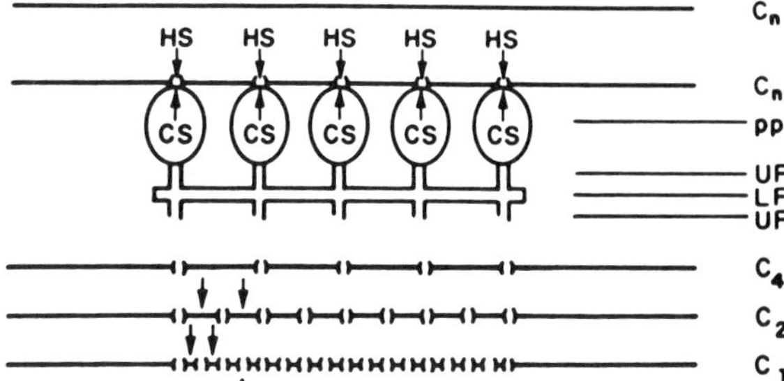

Fig. 2. Diagrammatic representation of the mode of action used by the cellulosome of *Clostridium thermocellum* to degrade cellulose.
Abbreviations: CS – catalytic site; HS – site of hydrolysis of cellulose; LF – central string of unknown material; C_n, C_4, C_2, C_1 – cellodextrins of various lengths. C_n is cellulose; C_1 is cellobiose. From: Mayer et al. 1987

Studies involving partial proteolysis have provided evidence of the possible functions of each do-

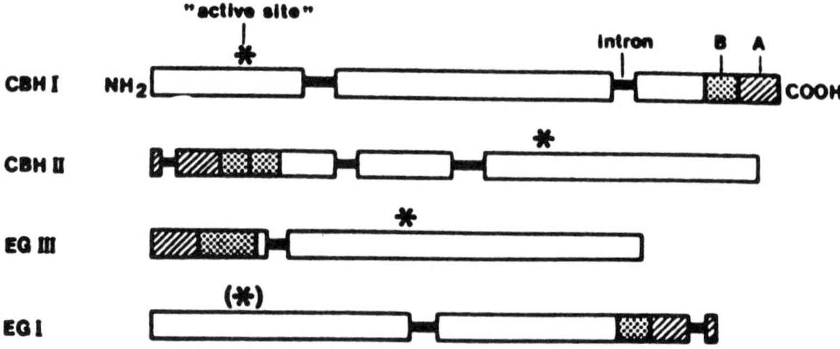

Fig. 3. Diagrammatic representation of four *Trichoderma* cellulase genes. Each enzyme comprises a tail domain (block *A*) linked to a flexible 'hinge' region (block *B*) to a catalytic domain (open boxes). The terminal domains are represented by striped boxes and the putative hinge regions by dotted boxes. A star indicates a putative active site. Intron positions are shown by solid bars.
From: Knowles et al. 1988b.

main. Thus, the conserved region appears to be involved in substrate binding and the core in catalysis. This was readily demonstrated in *T. reesei* CBH I and CBH II when it was found that removal of blocks A and B from both by limited proteolysis had a dramatic effect on the capacity of the core proteins of both CBH I and II to degrade microcrystalline cellulose (Tomme et al. 1988b). Indeed, the specific activity of the CBH I core was only 10% of that shown by the intact enzyme: CBH II core retained 40% of the original activity. However, the cores retained much of the capacity (93% in the case of CBH I core) of the intact enzyme to bind to amorphous cellulose: all the activity to small soluble cello-oligosaccharides was retained. From these data it was concluded that the tails are involved in binding the intact enzymes to the crystalline substrate.

The core protein released from *T. reesei* CBH I by proteolysis was 56 kDa, the smaller domain 10 kDa (van Tilbeurgh et al. 1986): the large and small domains in CBH II were 58 kDa and 13 kDa, respectively. Monoclonal antibodies raised against the respective cores of CBH I and II were highly specific (Mischak et al. 1989); no cross reactivities were observed.

Further studies carried out on the binding domain of *T. reesei* CBH I have established that the three dimensional structure of the binding peptide must be retained intact for biological function (Johansson et al. 1989). Using NMR spectroscopy, it has been shown that the binding domain has the shape of a wedge, with overall dimensions 30 × 18 × 10 Å (Teeri et al. 1990). One surface is flat and hydrophobic (Kraulis et al. 1989), the other is flat and hydrophilic. It has been postulated that this structure could interact with crystalline cellulose to effectively solubilize as a preliminary to hydrolysis by the core enzyme (Teeri et al. 1990).

Attention is now being directed at the core protein. Crystals of both CBH I and II cores have been obtained (Bergfors et al. 1989), but only the crystal structure of CBH II core protein has been elucidated (Rouvinen et al. 1989). The core is approximately 50 Å in diameter and consists of a seven-strand-

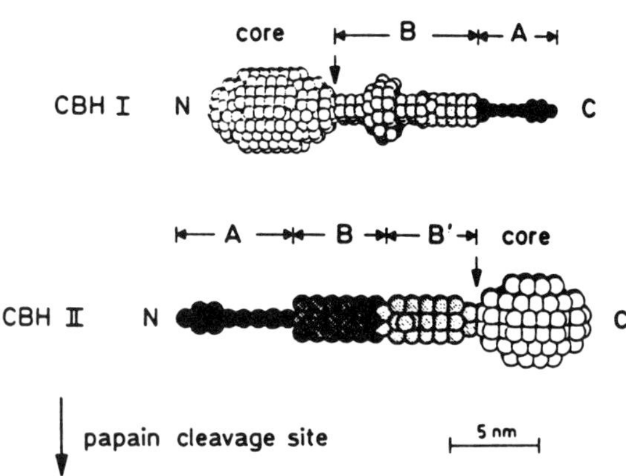

Fig. 4. Tertiary structure of CBH I and CBH II from *Trichoderma reesei* as deduced by small angle X-ray scattering. *A* and *B* relate to *A* and *B* in Fig. 3. From: Abuja et al. 1988b.

ed singly wound α-β-barrel. It seems that substrate binding occurs in a large channel which is formed by extended loops from the barrel (Teeri et al. 1990). The active site of the enzyme has been located at the COOH end of the β-sheets using an inhibitor diffused in the crystal (Teeri et al. 1990). It is not clear yet which amino acids constitute the active site but, as discussed below, progress has been made in this area.

Active site studies

Using chemical modification studies, Tomme & Claeyssens (1989) have identified Glu-126 to be a catalytically important carboxyl residue in CBH I of *T. reesei*: Glu-127 was proposed as a potential active site in EG I. A suggestion, based on limited homologies of the primary structures of the cellulases to the active sites of different lysozymes (Knowles et al. 1987), that Glu-65 and Asp-74 were involved in the active site was not supported by chemical modification studies. Glu-126 in *T. reesei* CBH I is reported to be located in a hydrophobic region between two large domains and to be equivalent to Glu-35 in egg white lysozyme.

Henrissat & Mornon (1990) have used hydrophobic cluster analysis to identify essential amino acids in a variety of cellulases. A histidyl residue appears to be involved in the active site of *C. thermocellum* (Claeyssens & Tomme 1989).

Stereochemical course of catalysis by cellulases

According to Reese et al. (1967), exoglucanases act by inversion of configuration. The question as to whether cellulases act by retention or inversion of anomeric configuration has been approached from time to time, as this has an important bearing on our conception of the mode of action of the cellobiohydrolases in particular. It now seems that this problem has been resolved. NMR spectroscopy of the hydrolysis of β-D-cellobiosylfluoride (Knowles et al. 1988a) suggests that the two cellobiohydrolases have different mechanisms: CBH I acts by retention of configuration while CBH II inverts. Similar conclusions have been reported by Claeyssens et al. (1990): CBH I of *T. reesei* liberated β-cellobiose from β-methyl cellobiose while CBH II, in contrast, liberated α-cellobiose from cello-oligosaccharides. EG I from *T. reesei*, which is structurally very similar to CBH I (see above), also acts by retention of configuration. Thus, while CBH I and EG I have a hydrolytic double inversion mechanism similar to that operating in lysozyme, CBH II may utilise a β-amylase type mechanism involving single displacement (Claeyssens & Tomme 1989). An exoglucanase from *C. fimi* acts by retention of anomeric configuration, while an endoglucanase from the same bacterium inverts (Withers et al. 1986).

Conclusions

Clearly, remarkable advances have been made in the last few years in the understanding of the cellulases and the study is in the middle of a very exciting phase. Further rapid advances in our understanding of the individual enzymes are in prospect now that,

- crystallisation of cellobiohydrolase cases has been possible and,
- attention is being directed at the development of an efficient host-vector system which will permit the reintroduction and expression of *Trichoderma* genes, modified by site specific mutagenesis, into cellulase negative mutants of *Trichoderma* (Teeri et al. 1990).

This will ensure effective enzyme processing and glycosylation and circumvent the problems associated with expression in a heterologous host. A study of the recombinant cellulase components which will be free from other contaminating cellulase components promises to provide unequivocal information on the hydrolytic properties of the individual enzymes.

Acknowledgements

The authors acknowledge the receipt of funding

from the Commission of the European Communities (contract MA1D 0014 UK [BA] and contract SC 1000205). They also wish to express their thanks to many colleagues who sent reprints and preprints of their work.

References

Abuja PM, Pilz I, Claeyssens M & Tomme P (1988a) Domain structure of cellobiohydrolase II as studied by small X-ray scattering: Close resemblance to cellobiohydrolase. I. Biochem. Biophys. Res. Commun. 156: 180–185

Abuja PM, Schmuck M, Pilz I, Tomme P & Claeyssens M et al. (1988b) Structural and functional domains of cellobiohydrolase I from *Trichoderma reesei*. A small angle X-ray scattering study of the intact enzyme and its core. Eur. Biophys. J. 15: 339–342

Bayer EA & Lamed R (1986) Ultrastructure of the cell surface cellulosome of *Clostridium thermocellum* and its interaction with cellulose. J. Bacteriol. 167: 828–836

Bayer EA, Setter E & Lamed R (1985) Organization and distribution of the cellulosome in *Clostridium thermocellum*. J. Bacteriol. 163: 552–559

Béguin P (1990) Molecular biology of cellulose degradation. Ann. Rev. Microbiol. 44: 219–248

Beldman G, Searle-Van Leewen MF, Rombouts FR & Voragen FGJ (1985) The cellulase of *Trichoderma viride*. Purification, characterization and comparison of all detectable endoglucanases, exoglucanases and β-glucosidases. Eur. J. Biochem. 146: 301–308

Bergfors T, Rouvinen J, Lethovaara P, Caldentey X, Tomme P, Claeyssens M, Pettersson G, Teeri T, Knowles J & Jones TA (1989) Crystallization of the core protein of cellobiohydrolase II from *Trichoderma reesei*. J. Molec. Biol. 209: 167–169

Bhat KM, McCrae SI & Wood TM (1989) The endo-1,4-β-D-glucanase system of *Penicillium pinophilum* cellulase: isolation, purification, and characterization of five major endoglucanase components. Carbohyd. Res. 190: 279–297

Bhikhabhai R & Pettersson G (1984) The disulphide bridges in a cellobiohydrolase and an endoglucanase from *Trichoderma reesei*. Biochem J. 222: 729–736

Changas GS & Wilson DB (1988) Cloning of the *Thermomonospora fusca* endoglucanase E2 gene in *Streptomyces lividans*: Affinity purification and functional domains of the cloned gene product. Appl. Environ. Microbiol. 54: 2521–2526

Chanzy H, Henrissat B & Vuong R (1984) Colloidal gold labelling of 1,4-β-D-glucan cellobiohydrolase adsorbed on cellulose substrates. FEBS Lett. 172: 193–197

Chanzy H & Henrissat B (1985) Unidirectional degradation of *Valonia* cellulose microcrystals subjected to cellulase action. FEBS Lett. 184: 285–288

Chanzy H, Henrissat B, Vuong R & Schülein M (1983) The action of 1,4-β-D-glucan cellobiohydrolase on *Valonia* cellulose microcrystals. An electron microscope study. FEBS Lett. 153: 113–118

Chen CM, Gritzali M & Stafford DW (1987) Nucleotide sequence and deduced primary structure of cellobiohydrolase II from *Trichoderma reesei*. Bio/Technol. 5: 274–278

Chippaux M (1988) Genetics of cellulase in *Erwinia chrysanthemi*. In: Aubert JP, Béguin P & Millet J (Eds) FEMS Symposium No. 43, Biochemistry and Genetics of Cellulose Degradation (pp 219–234). Academic Press, London

Claeyssens M, Van Tilbeurgh H, Tomme P, Wood TM & McCrae SI (1989) Comparison of the specificities of the cellobiohydrolases isolated from *Penicillium pinophilum* and *Trichoderma reesei*. Biochem. J. 261: 819–825

Claeyssens M & Tomme P (1989) Structure-activity relationships in cellulolytic enzymes. In: Coughlan MP (Ed) Enzyme Systems for Lignocellulose Degradation (pp 37–49). Elsevier Applied Science, London

Claeyssens M, Tomme P, Boewer CF & Hehre EJ (1990) Stereochemical course of hydrolysis and hydration reactions catalysed by cellobiohydrolases I and II from *Trichoderma reesei*. FEBS Lett. 263: 89–92

Coughlan MP (1985) The properties of fungal and bacterial cellulases with comment on their production and application. Biotechnol. Genet. Eng. Revs. 3: 39–109

(1990) Mechanisms of cellulose degradation by fungi and bacteria. In: Delort-Laval J (Eds) Cell Walls: Structure, Function and Degradation (in press)

Coughlan MP & Ljungdahl LG (1988) Comparative biochemistry of fungal and bacterial cellulolytic enzyme systems. In: Aubert JP, Béguin P & Millet J (Eds) FEMS Symposium No. 43, Biochemistry and Genetics of Cellulose Degradation (pp 11–30). Academic Press, London

Creuzet N, Berenger JF & Frixon C (1983) Characterization of exoglucanase and synergistic hydrolysis of cellulose in *Clostridium stercorarium*. FEMS Microbiol. Lett. 20: 347–350

Enari TM & Niku-Paavola ML (1987) Enzymatic hydrolysis of cellulose: Is the current theory of the mechanisms of hydrolysis valid? CRC Crit. Revs. Biotechnol. 5: 67–87

Eriksson KE & Pettersson B (1982) Purification and partial characterization of two acidic proteases from the white rot fungus *Sporotrichum pulverulentum*. Eur. J. Biochem. 124: 635–642

Eriksson KE & Wood TM (1985) Biodegradation of cellulose. In: Higuchi T (Ed) Biosynthesis and Biodegradation of Wood Components (pp 469–504). Academic Press

Fägerstam LG & Pettersson LG (1980) The 1,4-β-glucan cellobiohydrolases of *Trichoderma reesei* QM9414. FEBS Lett. 119: 97–101

Fägerstam LG, Pettersson LG & Engström JA (1984) The primary structure of a 1,4-β-glucan cellobiohydrolase from the fungus *Trichoderma reesei* QM9414. FEBS Lett. 167: 309–315

Faure E, Belaich A, Bagnara C, Gaudin C & Belaich JP (1989) Sequence analysis of the *Clostridium cellulolyticum* celCCA endoglucanase gene. Gene 65: 51–58

Gardner RM, Doewer KC & White BA (1987) Purification and

characterization of an exo-β-1,4-glucanase from *Ruminococcus flavefaciens* FD-1. J. Bacteriol. 169: 4581–4588

Gilkes NR, Warren RAJ, Miller RC Jr & Kilburn DG (1988) Precise excision of the cellulose binding domains from two *Cellulomonas fimi* cellulases by a homologous protease and the effect on catalysis. J. Biol. Chem. 263: 10401–10407

Gräbnitz F & Staudenbauer WL (1988) Characterization of two β-glucosidase genes from *Clostridium thermocellum*. Biotechnol. Lett. 10: 73–78

Gum EK & Brown RD (1977) Comparison of four purified extracellular 1,4-β-D-glucan cellobiohydrolase enzymes from *Trichoderma viride*. Biochim. Biophys. Acta 492: 225–231

Hall J, Hazlewood GP, Barker PJ & Gilbert HJ (1988) Conserved reiterated domains in *Clostridium thermocellum* endoglucanases are not essential for activity. Gene 69: 29–38

Hazlewood GP, Romaniec MP, Davidson K, Grépinet O & Béguin P (1988) A catalogue of *Clostridium thermocellum* endoglucanase, β-glucosidase and xylanase genes cloned in *Escherichia coli*. FEMS Microbiol. Lett. 51: 231–236

Henrissat B, Claeyssens M, Tomme P, Lemesle L & Mornon JP (1989) Cellulase families revealed by hydrophobic cluster analysis. Gene 81: 83–95

Henrissat B, Driguez H, Viet C & Schülein M (1985) Synergism of cellulases from *Trichoderma reesei* in the degradation of cellulose. Bio/Technol. 3: 722–726

Henrissat B & Mornon JP (1990) Comparison of *Trichoderma* cellulases with other β-glucanases. In: Kubicek CP, Eveleigh DE, Eisterbauer H, Steiner W & Kubicek-Pranz EM (Eds) *Trichoderma reesei* Cellulases: Biochemistry, Genetics, Physiology, and Applications. Proc. TRICEL 89 Meet. held in Vienna, Sept. 1989. Royal Society of Chemistry

Howard GT & White B (1988) Molecular cloning and expression of cellulase genes from *Ruminococcus albus* 8 in *Escherichia coli* bacteriophage. Appl. Environ. Microbiol. 54: 1752–1755

Johansson G, Ståhlberg J, Lindeberg G, Engström Å & Pettersson G (1989) Isolated fungal cellulase terminal domains and a synthetic minimum analogue bind to cellulose. FEBS Lett. 243: 389–393

Johnson EA, Sakajah M, Halliwell G, Madia A & Demain AL (1982) Saccharification of complex cellulosic substrates by the cellulase system from *Clostridium thermocellum*. Appl. Environ. Microbiol. 43: 1125–1132

Joliff G, Béguin P, Juy M, Millet J & Ryter A (1986) Isolation, crystallization and properties of a new cellulase of *Clostridium thermocellum* overproduced in *Escherichia coli*. Bio/Technol. 4: 896–900

Klyosov AA (1988) Cellulases of the third generation. In: Aubert JP, Béguin P & Millet J (Eds) FEMS Symposium No. 43, Biochemistry and Genetics of Cellulose Degradation (pp 87–99). Academic Press, London

Knowles JKC, Lehtovaara P & Teeri TT (1987) Cellulase families and their genes. Trends Biotechnol. 5: 255–261

Knowles JKC, Lehtovaara P, Murray M & Sinnott M (1988a) Stereochemical course of action of the cellobioside hydrolases I and II of *Trichoderma reesei*. J. Chem. Soc., Chem. Commun. 1401–1402

Knowles JKC, Teeri TT, Lehtovaara P, Penttilä M & Saloheimo M (1988b) The use of gene technology to investigate fungal cellulolytic enzymes. In: Aubert JP, Béguin P & Millet J (Eds) FEMS Symposium No. 43, Biochemistry and Genetics of Cellulose Degradation (pp 153–169). Academic Press, London

Koenigs JW (1975) Hydrogen peroxide and iron: A microbial cellulolytic system. Biotechnol. Bioeng. Symp. 5: 151–159

Kraulis PJ, Clore GM, Nilges M, Jones TA, Pettersson G, Knowles JKC & Gronenborn AM (1989) Determination of the three dimensional structure of the C-terminal domain of cellobiohydrolase I from *Trichoderma reesei*. A study using nuclear magnetic resonance and hybrid distance geometry-dynamical simulated annealing. Biochemistry 28: 7241–7257

Kyriacou AK, MacKenzie CR & Neufield RJ (1987) Detection and characterization of specific and non-specific endoglucanases of *Trichoderma reesei*. Evidence demonstrating endoglucanase activity by cellobiohydrolase II. Enzyme Microb. Technol. 9: 25–32

Lamed R & Bayer EA (1988) The cellulosome concept: Exocellular/extracellular enzyme reactor centers for efficient binding and cellulolysis. In: Aubert JP, Béguin P & Millet J (Eds) FEMS Symposium No. 43, Biochemistry and Genetics of Cellulose Degradation (pp 101–116). Academic Press, London

Lamed R, Setter E & Bayer EA (1983a) Characterization of a cellulose-binding, cellulase-containing complex in *Clostridium thermocellum*. J. Bacteriol. 156: 828–836

Lamed R, Setter E, Kenig R & Bayer EA (1983b) The cellulosome: A discrete cell surface organelle of *Clostridium thermocellum* which exhibits separate antigenic, cellulose-binding and various cellulolytic activities. Biotechnol. Bioeng. Symp. 13: 163–181

Lamed R, Kenig R & Setter E (1985) Major characteristics of the cellulolytic system of *Clostridium thermocellum* coincide with those of the purified cellulosome. Enzyme Microb. Technol. 7: 32–41

Ljungdahl LG (1989) Mechanisms of cellulose hydrolysis by enzymes from anaerobic and aerobic bacteria: In: Coughlan MP (Ed) Enzyme Systems for Lignocellulose Degradation (pp 5–16). Elsevier Applied Science, London

Ljungdahl LG, Coughlan MP, Mayer F, Mori Y & Hon-nami K (1988) Macrocellulase complexes and yellow affinity substance from *Clostridium thermocellum*. In: Wood WA & Kellog ST (Eds) Methods in Enzymology, Vol 160 (pp 483–500). Academic Press, New York

Ljungdahl LG, Petterson B, Eriksson KE & Wiegel J (1983) A yellow affinity substance involved in the cellulolytic system of *Clostridium thermocellum*. Curr. Microbiol. 9: 195–200

McGavin M & Forsberg CW (1989) Catalytic and substrate-binding domains of endoglucanase 2 from *Bacteroides succinogenes*. J. Bacteriol. 121: 3310–3315

MacKenzie CR, Bilous D & Johnson KG (1984) Purification

and characterization of an exoglucanase from *Streptomyces flavogriseus*. Can. J. Microbiol. 30: 1171–1178

MacKenzie CR, Bilous D & Patel GB (1985) Studies on cellulose hydrolysis by *Acetivibrio cellulolyticus*. Appl. Environ. Microbiol. 50: 243–248

MacKenzie CR, Patel GB & Bilous D (1987) Factors involved in hydrolysis of microcrystalline cellulose by *Acetivibrio cellulolyticus*. Appl. Environ. Microbiol. 53: 304–308

Mayer F (1988) Cellulolysis: Ultrastructural aspects of bacterial systems. Electron Microsc. Rev. 1: 69–85

Mayer F, Coughlan MP, Mori Y & Ljungdahl LG (1987) Macromolecular organization of the cellulolytic enzyme complex of *Clostridium thermocellum* as revealed by electron microscopy. Appl. Environ. Microbiol. 53: 2785–2792

Miller RC Jr, Gilkes NR, Greenberg NM, Kilburn DG, Langsford ML & Warren RAJ (1988) *Cellulomonas fimi* cellulases and their genes. In: Aubert JP, Béguin P & Millet J (Eds) FEMS Symposium No. 43, Biochemistry and Genetics of Cellulose Degradation (pp 235–248). Academic Press, London

Mischak H, Hofer F, Messner R, Weissinger E & Hayn M (1989) Monoclonal antibodies against different domains of cellobiohydrolase I and II from *Trichoderma reesei*. Biochim. Biophys. Acta 990: 1–7

Niku-Paavola ML, Lappalainen A, Enari TM & Nummi M (1985) A new appraisal of the endoglucanases of the fungus *Trichoderma reesei*. Biochem. J. 231: 75–81

Penttilä M, Lehtovaara P, Nevalainen H, Bhikhabhai R & Knowles JKC (1986) Homology between cellulase genes of *Trichoderma reesei*: Complete nucleotide sequence of the endoglucanase I gene. Gene 45: 253–263

Pilz I, Schwarz E, Kilburn DG, Miller RC Jr, Warren RAJ & Gilkes NR (1990) The tertiary structure of a bacterial cellulase determined by small-angle X-ray-scattering analysis. Biochem. J. 271: (in press)

Reese ET, McGuire AH & Parrish FW (1967) Glucosidase and exo-glucanases. Can. J. Biochem. 46: 25–34

Reese ET, Siu RGH & Levinson HS (1950) Biological degradation of soluble cellulose derivatives. J. Bacteriol. 9: 485–497

Rouvinen J, Bergfors T, Pettersson G, Knowles JKC & Jones TA (1989) Crystallographic studies on the core protein of cellobiohydrolase II from *Trichoderma reesei*. First European Workshop on Crystallography of Biological Macromolecules. Como, Italy, May 15–19, 1989

Ryu DDY, Kim C & Mandels M (1984) Competitive adsorption of cellulase components and its significance in a synergistic mechanism. Biotechnol. Bioeng. 26: 488–496

Saloheimo M, Lehtovaara P, Penttilä M, Teeri TT & Stahlberg J (1988) EGIII, a new endoglucanase from *Trichoderma reesei*: and the characterization of both gene and enzyme. Gene 63: 11–21

Schmuck M, Pilz I, Hayn M & Esterbauer H (1986) Investigation of cellobiohydrolase from *Trichoderma reesei* by small angle X-ray scattering. Biotechnol. Lett. 8: 397–402

Shoemaker S, Schweickaert V, Ladner M, Gelfand D, Kwok S, Myambo K & Innis M (1983) Molecular cloning of exo-cellobiohydrolase I derived from *Trichoderma reesei* strain L27. Bio/Technol. 1: 691–696

Sprey B & Lambert C (1983) Titration curves of cellulases from *Trichoderma reesei*: Demonstration of a cellulase-xylanase-β-glucosidase-containing complex. FEMS Microbiol. Lett. 18: 217–222

Streamer M, Eriksson KE & Pettersson B (1975) Extracellular enzyme system utilized by the fungus *Sporotrichum pulverulentum* (*Chrysosporium lignorum*) for the breakdown of cellulose. Functional characterization of five endo-1,4-β-glucanase and one exo-β-1,4-glucanase. Eur. J. Biochem. 59: 607–613

Teeri TT, Jones A, Kraulis P, Rouvinen J, Penttilä M, Harkki A, Nevelainen H, Vanhanen S, Saloheimo M & Knowles JKC (1990) Engineering *Trichoderma* and its cellulases. In: Kubicek CP, Eveleigh DE, Esterbauer H, Steiner W & Kubicek-Pranz EM (Eds) *Trichoderma reesei* Cellulases: Biochemistry, Genetics, Physiology, and Applications. Proc. TRICEL 89 Meet. held in Vienna, Sept. 1989. Royal Society of Chemistry

Teeri TT, Kumar V, Lehtovaara P & Knowles JKC (1987a) Construction of cDNA libraries by blunt end ligation: high-frequency cloning of long cDNAs from filamentous fungi. Anal. Biochem. 164: 60–67

Teeri TT, Lehtovaara S, Kauppinen S, Salovuori I & Knowles JKC (1987b) Homologous domains in *Trichoderma reesei* cellulolytic enzymes: gene, sequence and expression of cellobiohydrolase II. Gene 51: 43–52

Teeri TT, Salovouri I & Knowles JKC (1983) The molecular cloning of the major cellulase gene from *Trichoderma reesei*. Bio/Technol. 1: 696–699

Tomme P & Claeyssens M (1989) Identification of a functionally important carboxyl group in cellobiohydrolase I from *Trichoderma reesei*: A chemical modification study. FEBS Lett. 243: 239–243

Tomme P, Heriban V & Claeyssens M (1990) Adsorption of two cellobiohydrolases from *Trichoderma reesei* to Avicel: evidence for 'exo-exo' synergism and possible 'loose complex' formation. Biotech. Lett. 121: 525–530

Tomme P, McCrae SI, Wood TM & Claeyssens M (1988a) Chromatographic separation of cellulolytic enzymes. In: Wood WA & Kellog ST (Eds) Methods in Enzymology. Vol 160 (pp 187–193). Academic Press, New York

Tomme P, Van Tilbeurgh H, Pettersson G, Van Damme J, Vandekerckhove J, Knowles JKC, Teeri TT & Claeyssens M (1988b) Studies of the cellulolytic system of *Trichoderma reesei* QM 9414. Eur. J. Biochem. 170: 575–581

Van Arsdell JN, Kwok S, Schweickart VL, Ladner MB, Gelfand DH & Innis MA (1987) Cloning, characterization and expression in *Saccharomyces cerevisiae* of endoglucanase I from *Trichoderma reesei*. Bio/Technol. 4: 60–64

Van Tilbeurgh H, Claeyssens M & de Bruyne CK (1982) The use of 4–methylumbelliferyl and other chromophoric glycosides in the study of cellulolytic enzymes. FEBS Lett. 149: 152–156

Van Tilbeurgh H, Bhikhabhai R, Pettersson LG & Claeyssens

M (1984) Separation of endo- and exo-type cellulases using a new affinity chromatography method. FEBS Lett. 169: 215–218

Van Tilbeurgh H, Tomme P, Claeyssens M, Bhikhabhai R & Pettersson G (1986) Limited proteolysis of the cellobiohydrolase I from *Trichoderma reesei*. FEBS Lett. 204: 223–227

Warren RAJ, Beck CF, Gilkes NR, Kilburn DG & Langsford M (1986) Sequence conservation and region shuffling in an endoglucanase and an exoglucanase from *Cellulomonas fimi*. Proteins Struct. Funct. Genet. 1: 335–341

White AR & Brown RM (1981) Enzymatic hydrolysis of cellulose: Visual characterization of the process. Proc. Nat. Acad. Sci. U.S.A. 78: 1047–1051

Wilters SG, Dombroski D, Beaven LA, Kilburn DG, Miller RC Jr, Warren RAJ & Gilkes NR (1986) Direct ^{1}H NMR determination of the stereochemical course of hydrolysis catalyzed by glucanase components of the cellulase complex. Biochem. Biophys. Res. Commun. 139: 487–494

Wood TM (1975) Properties and mode of action of cellulases. Biotechnol. Bioeng. Symp. 5: 111–137

(1989) Mechanisms of cellulose degradation by enzymes from aerobic and anaerobic fungi. In: Coughlan MP (Ed) Enzyme Systems in Lignocellulose Degradation (pp 17–35). Elsevier Applied Science, London

(1990) Fungal cellulases. In: Weimer PJ & Hagler CA (Eds) Biosynthesis and Biodegradation of Cellulose and Cellulosic Materials. Marcel Dekker, New York (in press)

Wood TM & McCrae SI (1972) The purification and properties of the C1 component of *Trichoderma koningii* cellulase. Biochem. J. 128: 1183–1192

(1979) Synergism between enzymes involved in the solubilization of native cellulose. Adv. Chem. Ser. 181: 181–209

(1986a) Purification and properties of a cellobiohydrolase from *Penicillium pinophilum*. Carbohyd. Res. 148: 331–334

(1986b) The cellulase of *Penicillium pinophilum*. Synergism between enzyme components in solubilizing cellulose with special reference to the involvement of two immunologically-distinct cellobiohydrolases. Biochem. J. 234: 93–99

Wood TM, McCrae SI & Bhat KM (1989) The mechanism of fungal cellulase action. Synergism between enzyme components of *Penicillium pinophilum* cellulase in solubilizing hydrogen bond-ordered cellulose. Biochem. J. 260: 37–43

Wood TM, McCrae SI & MacFarlane CC (1980) The isolation, purification and properties of the cellobiohydrolase component of *Penicillium funiculosum* cellulase. Biochem. J. 189: 51–65

Wood TM, McCrae SI, Wilson CA, Bhat KM & Gow LA (1988) Aerobic and anaerobic fungal cellulases, with special reference to their mode of attack on crystalline cellulose. In: Aubert JP, Béguin P & Millet J (Eds) FEMS Symposium No. 43, Biochemistry and Genetics of Cellulose Degradation (pp 31–52). Academic Press, London

Woodward J, Hayes MK & Lee NL (1988b) Hydrolysis of cellulose by saturating and non-saturating concentration of cellulase: implications for synergism. Bio/Technol. 6: 301–304

Woodward J, Lima M & Lee NL (1988a) The rôle of cellulase concentration in determining the degree of synergism in the hydrolysis of microcrystalline cellulose. Biochem. J. 255: 895–899

Wu JDH, Orme-Johnson WH & Demain AL (1988) Two components of an extracellular protein aggregate of *Clostridium thermocellum* together degrade crystalline cellulose. Biochem. 27: 1703–1709

Yablonsky MD, Bartley T, Elliston KO, Kahrs SK, Shalita ZP & Eveleigh DE (1988) Characterization and cloning of the cellulase complex of *Microbispora bispora*. In: Aubert JP, Béguin P & Millet J (Eds) FEMS Symposium No. 43, Biochemistry and Genetics of Cellulose Degradation (pp 249–266). Academic Press, London

Yablonsky MD, Elliston KO & Eveleigh DE (1989) The relationship between the endoglucanase gene MbcelA of *Microbispora bispora* and cellulase genes of *Cellulomonas fimi*. In: Coughlan MP (Ed) Enzyme System for Lignocellulose Degradation (pp 73–83). Elsevier Applied Science, London

Biodegradation **1**: 163–176, 1990.

Biodegradation of lignin-carbohydrate complexes

Thomas W. Jeffries
Institute for Microbial and Biochemical Technology, USDA Forest Service, Forest Products Laboratory, One Gifford Pinchot Drive, Madison, WI 53705-2398 U.S.A.

Key words: bonds, carbohydrate, lignin, enzymatic degradation

Abstract

Covalent lignin-carbohydrate (LC) linkages exist in lignocellulose from wood and groups herbaceous plants. In wood, they consist of ester and ether linkages through sugar hydroxyl to the α-carbanol of phenylpropane subunits in lignin. In grasses, ferulic and *p*-coumaric acids are esterified to hemicelluloses and lignin, respectively. Hemicelluloses also contain substituents and side groups that restrict enzymatic attack. Water-soluble lignin-carbohydrate complexes (LCCs) often precipitate during digestion with polysaccharidases, and the residual sugars are more diverse than the bulk hemicellulose. A number of microbial esterases and hemicellulose polysaccharidases including acetyl xylan esterase, ferulic acid esterase, and *p*-coumaric esterase attack hemicellulose side chains. Accessory hemicellulases include α-L-arabinofuranosidase and α-methyl-glucuranosidase. Both of these side chains are involved in LC bonds. β-Glucosidase will attach sugar residues to lignin degradation products and when carbohydrate is attached to lignin, lignin peroxidase will depolymerize the lignin more readily.

Abbreviations: APPL – acid precipitable polymeric lignin; CBQase – cellobioquinone oxidoreductase; LC – lignincarbohydrate; LCC(s) – lignin-carbohydrate complex; DHP – Dehydrogenative polymerisate; DMSO – dimethylsulfoxide; DP – degree of polymerisation; MWEL – milled wood enzyme lignin; MWL – milled wood lignin (not digested with carbohydrases)

Introduction

This review explores the characteristics and biodegradation of bonds between lignin and carbohydrate. Lignin-carbohydrate complexes (LCCs) can be isolated as water-soluble entities from the walls of gymnosperms, angiosperms, and graminaceous plants (Azuma & Koshijima 1988), and they can be separated by gel filtration into three fractions. The component of lowest molecular weight consists mostly of carbohydrate, the two larger components mostly of lignin. Lignin-carbohydrate bonds are presumed to exist in higher molecular weight lignin fractions that are water insoluble. Softwood LCCs are distinct in that their carbohydrate portions consist of galactomannan, arabino-4-*O*-methylglucuronoxylan, and arabinogalactan linked to lignin at benzyl positions (Azuma et al. 1981; Mukoyoshi et al. 1981). In contrast, carbohydrate portions of hardwood and grass LCCs are composed exclusively of 4-*O*-methylglucuronoxylan and arabino-4-*O*-methylglucuronoxylan, respectively (Azuma & Koshijima 1988). *Trans-p*-coumaric and *p*-hydroxybenzoic acid are esterified to bamboo and poplar lignins, respectively (Shimada et al. 1971), and *trans*-ferulic acid is ether linked to lignin (Scalbert et al. 1985). Many different types of LC bonds have been proposed, but most evidence exists for ether and ester linkages.

Relatively little attention has been given to en-

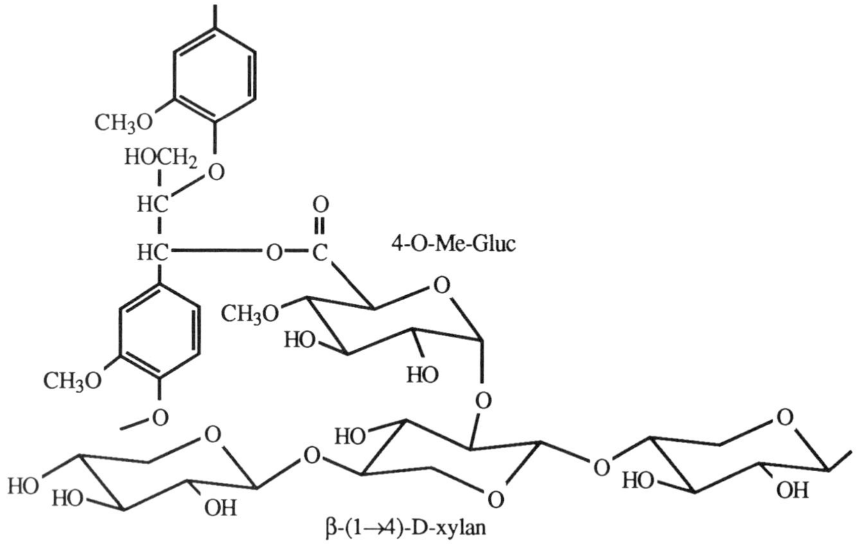

Fig. 1. Proposed structure of ester linkage between lignin and 4-*O*-methylglucuronoxylan in pine (after Watanabe & Koshijima 1988)

zymes capable of cleaving the chemical linkages between lignin and carbohydrate, the LC bonds. Such bonds occur in low frequency. They are heterogeneous; many are easily disrupted by acid or alkali during isolation, and they are still poorly defined. Although many sorts of linkages have been proposed, two have some substantive evidence. They link the α position of the phenyl propane lignin moiety to either carboxyl or free hydroxyls of hemicellulose through ester or ether linkages, respectively. Various chemical and enzymatic procedures have been used to isolate LC complexes, and a few biological systems have been shown to solubilize lignin preparations. No enzymes specific to LC bond cleavage have been described. The objective of this review is to focus our knowledge of the heterogeneous structures that comprise LC bonds, and to sort out the enzymes that attack related structures. A recent review of LCCs has been completed by Koshijima et al. (1989).

Chemical characteristics of LC bonds

Ester linkages (CO-O-C) occur between the free carboxy group of uronic acids in hemicellulose and the benzyl groups in lignin. Some are present as acetyl side groups on hemicellulose, others are between uronic acids and lignin, and still others occur between hemicellulose chains. Monomeric side chains in wood xylans consist of 4-*O*-methylglucuronic acid units, and some 40% of the uronic acid groups in birch are esterified. In beech, one-third of the glucuronic acids present in LCCs are involved in an ester linkage between lignin and glucuronoxylan (Takahishi & Koshijima 1988b). However, many glucuronic acid groups may be esterified within the xylan polymer (Wang et al. 1967).

Direct evidence for the chemical nature of ester linkages between lignin and carbohydrate in pine has been obtained through the selective oxidation of carbonyls in lignin. Watanabe & Koshijima (1988) proposed that the 4-*O*-methylglucuronic acid residue in arabinoglucuronoxylan binds to lignin by an ester linkage in *Pinus densiflora* wood. The linkage position is probably the α or conjugated γ position of guaiacylalkane units (Fig. 1). Watanabe et al. (1989) found that mannose, galactose, and glucose are *O*-6 ether linked and xylose is *O*-2 or *O*-3 ether linked to the a benzyl hydroxyl in a neutral fraction of pine LCC.

Watanabe et al. (1989) also studied alkali-stable LC linkages. The alkali-stable linkages in LCC prepared from *Pinus densiflora* consist of acetyl glucomannan and β-(1→ 4) galactan bound to the

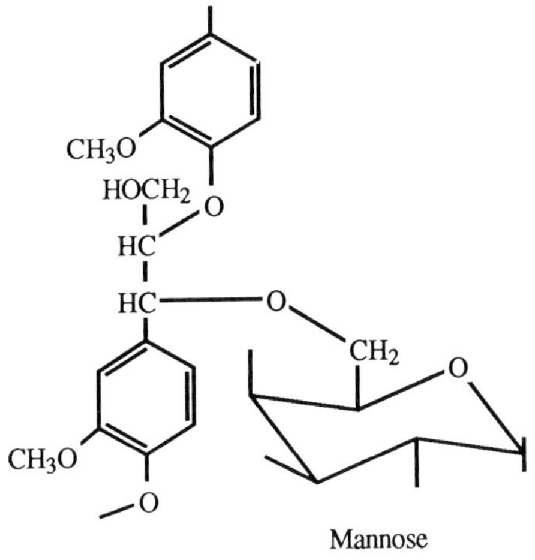

Fig. 2. Proposed structure for ether linkages between lignin and glucomannan of pine (after Watanabe et al. 1989).

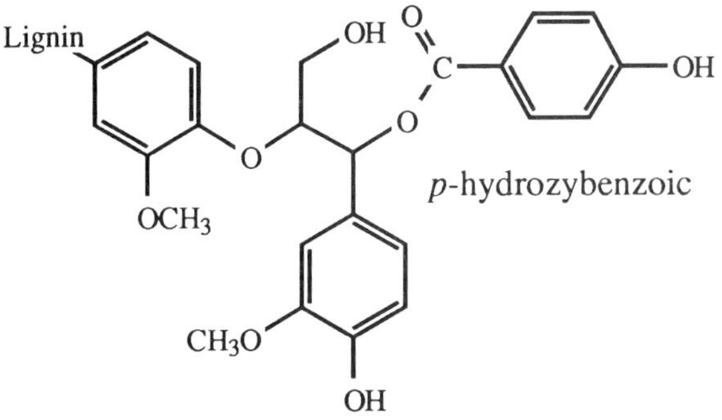

lignin at the *O*-6 position of the hexoses (Fig. 2). The GC-MS analysis of methylated sugar derivatives led Watanabe et al. (1989) to conclude that the arabinoglucuronoxylan is bound to the lignin at the *O*-2 and *O*-3 positions of the xylose units, and that the linkage position in the lignin subunits is in the α or conjugated β positions of phenyl propane or propene units.

Minor (1982) investigated the LC bonds of loblolly pine MWEL to determine the positions of linkages in carbohydrates. The carbohydrates exist as oligomeric chains with degrees of polymerization of 7 to 14. Hexose units are bonded at *O*-6; L-arabinose is bonded exclusively at *O*-5. Galactan and arabinan are structurally of the 1→4 and 1→5 type, respectively, characteristic of the neutral substituents of pectins. In spruce, lignin is linked by ester bonds to 4-*O*-methylglucuronic acid, and arabinoxylan is linked through ether bonds (C-O-C) to the *O*-2 or *O*-3 positions of L-arabinose (Ericksson et al. 1980). For galactoglucomannan, ether bonds to position 3 of galactose have been indicated.

Esters of *p*-coumaric acid in sugar cane lignin and of *p*-hydroxybenzoic acid in aspen lignins were first demonstrated by Smith (1955) who proposed that the carboxy group of *p*-hydroxybenzoic acid is esterified to lignin in the α-position of the phenyl propane (Fig. 3).

Pectic substances may play an important role in binding lignin to the hemicellulose. Pectins are able to form both ester and ether linkages with lignin. LCC from birch contains about 7% galacturonic and 4% glucuronic acid, and a small amount of galacturonic acid is found in the LCC of spruce (Meshitsuka et al. 1983). Pectins are abundant in some fiberous plant materials such as mitsumata (*Edgeworthia papyrifera*). Fiber bundles are held together by pectic substances, and the pectins are aggregated with LCCs. Aggregates between lignin and pectins are particularly present in bast fibers, and *endo*-pectin and *endo*-pectate lyases from the soft-rot bacterium *Erwinia carotovora* release pectic fragments from this substrate (Tanabe & Kobayashi 1988). Alkaline presoaking accelerates biochemical pulping of mitsumata by pectinolytic enzymes (Tanabe & Kobayashi 1986, 1987).

Fig. 3. Proposed structure for lignin-*p*-hydroxybenzoic acid ester in aspen.

Esterified *p*-coumaric acid can comprise 5% to 10% of the total weight of isolated grass or bamboo lignin (Shimada et al. 1971), but *p*-hydroxyphenyl glycerol-β-aryl ether structures are of minor importance (Higuchi et al. 1967). The majority of *p*-coumaric acid molecules in bamboo and grass lignins are ester-linked to the terminal γ carbon of the side chain of the lignin molecule (Shimada et al. 1971) (Fig. 4). The *p*-coumaric ester linkages are extremely stable.

p-Coumaric and ferulic acids are bifunctional, they are able to form ester or ether linkages by reaction of their carboxyl or phenolic groups, respectively. *p*-Coumaric is mainly associated with lignin; ferulic acid, on the other hand, is mainly esterified with hemicellulose (Scalbert et al. 1985;

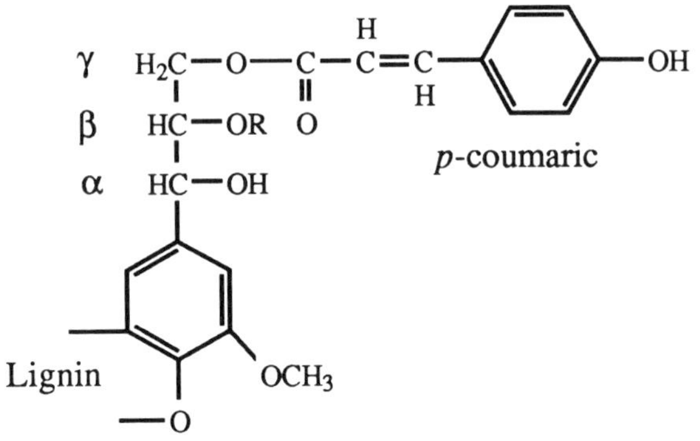

Fig. 4. Proposed *p*-coumaric ester linkage in grass lignins (after Shimada et al. 1971).

Atushi et al. 1984). Ferulic acid ethers might form cross links between lignin and hemicelluloses by the simultaneous esterification of their carboxyl group to arabinose substituents of arabinoglucuronxylan and etherification of their hydroxyl group to phenyl hydroxyls of lignin. Diferulic acids can be formed by polymerization with peroxidase/water, and studies suggest that these acids can crosslink hemicellulosic chains (Markwalder & Neukom 1976; Morrison 1974) (Fig. 5). Feruloylated arabinoxylans have been isolated from the LCC of bagasse (Kato et al. 1987), bamboo shoot cell walls (Ishii & Hiroi 1990), and pangola grass (*Digitaria decumber*) (Ford 1989) (Fig. 6).

The amount of carbohydrate remaining on lignin following exhaustive digestion with cellulases and hemicellulases can be measured by sugar analysis following acid hydrolysis. Obst (1982) found 10.8% carbohydrate in a MWEL from loblolly pine. A fraction of this amount (11%) was removed by dilute alkali. Obst estimated that there are approximately three LC bonds for every 100 phenylpropane units in lignin. The length of the side chains varies with the method by which MWEL is prepared. If extensive alkaline hydrolysis is performed prior to enzymatic digestion, a larger fraction of the residual carbohydrate is removed.

Biodegradation of LCCs

Enzymatic digestion of LCCs

Most carbohydrate chains or side groups appear to be attached to lignin through the non-reducing moieties. Because *exo*-splitting enzymes generally attack a substrate from the nonreducing end of a polysaccharide, removing substituents progressively toward the reducing end of the molecule, complete degradation is not possible. Even when carbohydrates are attached to the lignin by the *O*-1 hydroxyl, a single sugar residue could remain attached even after complete attack by *exo*-splitting glycosidases.

The action of *endo*-splitting glycanases is even more constrained. The binding sites of most *endo*-splitting polysaccharidases have not yet been well characterized. However, from transferase activities and other kinetic studies, Biely et al. (1981) showed that the substrate binding site of the *endo*-xylanase from *Cryptococcus* has eight to ten subsites for binding pyranose rings and the catalytic groups that make up the active site are located in the center. It is not surprising, therefore, that digestion of LCCs with *endo*-xylanases and *endo*-cellulases leaves residual polysaccharide oligomers with a degree of polymerization (DP) of 4 or more attached to the lignin moiety.

The action of an *endo*-xylanase purified from a commercial preparation from *Myrothecium verrucaria* illustrates the effects of substrate binding (Comtat et al. 1974). This preparation will cleave xylose residues from the nonreducing end of aspen 4-*O*-methylglucuronoxylan until it leaves a 4-*O*-methyl-α-D-glucuronic acid substituent linked 1→2 to a β-(1→4)-xylotriose. The enzyme cannot cleave the two β-(1→4)-D-xylan linkages immediately to the right of the 4-*O*-methylglucuronic acid side chain (in the direction of the reducing end). The residual xylotriose represents that portion of the substrate that is bound but that the enzyme cannot cleave. Presumably, other *endo*-splitting enzymes encounter even more extreme difficulties when carbohydrate is chemically linked adjacent to the lignin polymer because the side chains are generally greater than four residues. This could also be

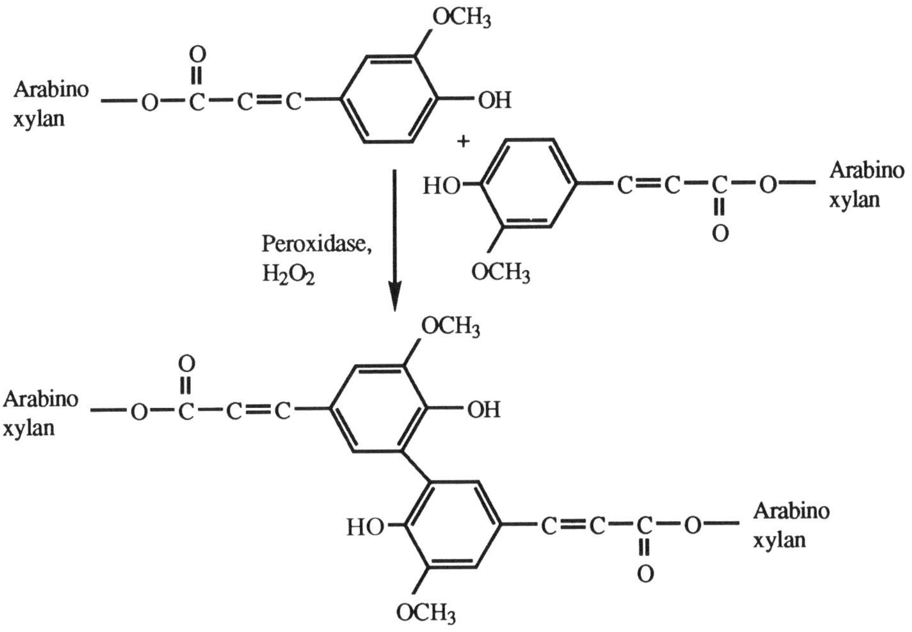

Fig. 5. Formation of diferulic acid in grasses (after Markwalder & Neukom 1975).

attributable to steric hindrance or to the presence of multiple cross-links between the lignin and carbohydrate polymers.

If there is low enzymatic activity for a particular side chain, longer hemicellulosic branches are produced. This was evidenced with the apparent enrichment of galactomannan during enzymatic digestion of Norway spruce (Iversen et al. 1987). The low activity of β-(1→ 4)-D-galactanase in enzyme preparations from *Trichoderma reesei* and *Aspergillus niger* led to the enrichment of galactomannan in the MWEL. One must be on guard against these possibilities when studying the structures of lignin complexes prepared by enzymatic digestion.

Through selective enzymatic degradation, it should be possible to identify which carbohydrates form LC linkages. Joseleau & Gancet (1981) took such an approach in characterizing residual carbohydrate after enzymatic digestion, alkaline hydrolysis (0.5 M NaOH), and mild acid hydrolysis (0.01 M oxalic) of an LCC isolated from aspen by dimethylsulphoxide (DMSO) extraction. The original LCC contained mostly carbohydrate, and the final material was predominantly lignin. Enzyme treatments and alkali extraction removed the bulk of the xylose and glucose while increasing the relative portions of arabinose and galactose. Rhamnose was also very persistent; its fraction relative to other sugars increased appreciably, indicating that rhamnose might be involved in an LC bond. Uronic acid likewise was significantly represented in the enzyme-, alkali-, and acid-treated material. It is not known whether uronic acid remains associated because of cross-linkages between lignin and carbohydrate or simply because of the stability of glycosidic linkages between uronic acids and xylan.

Residual LC structures after exhaustive enzymatic digestion

The presence of lignin, aromatic acids, and other modifications of hemicellulose clearly retards digestion of cellulose and hemicellulose by ruminants. Phenolic acids associated with forage fiber are known to reduce fiber digestion when they are in the free state. *p*-Coumaric, ferulic, and sinapic acids inhibit the activity of rumen bacteria and anaerobic fungi (Akin & Rigsby 1985). It seems likely, however, that the cross-linkages that these acids establish between hemicellulose chains and the lignin polymer are more important than is the

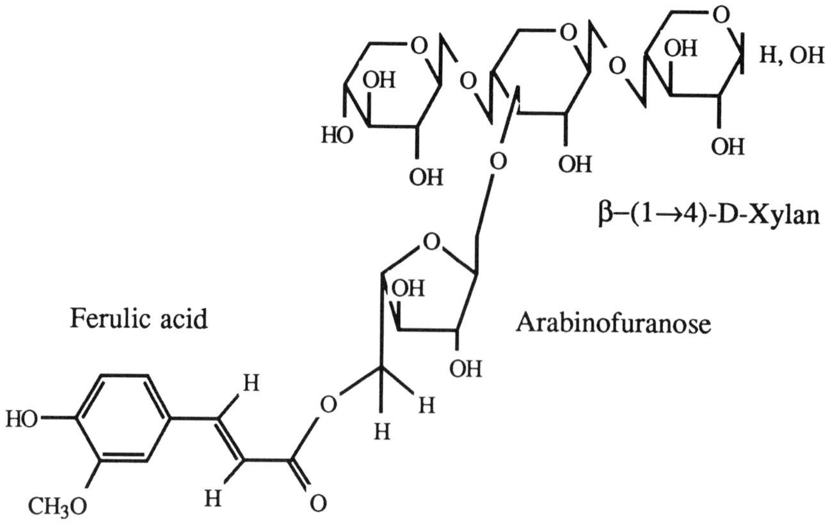

Fig. 6. Ferulic acid ester linkage to grass arabinoxylan (after Kato et al. 1987).

toxicity attributable to the monomers themselves because ruminant digestibility is greatly enhanced by alkali treatment, regardless of whether the phenolic acids are washed out of the residue.

In a comparison of different ruminant feeds, an increase in the extent of substitution at the *O*-5 position of arabinose correlated with the amount of residual material in digested residues (Chesson et al. 1983). The influence of the lignin, however, was out of proportion to its presence in the feed (Chesson 1988). When a steer digests grass, LCCs are solubilized in the rumen. This dissolution accounts for about half the total lignin intake (Gaillard & Richards 1975). At least two studies of LCCs isolated from ruminants have been published (Neilson & Richards 1982; Conchie et al. 1988). The amounts of residual carbohydrate in the LCC isolated from ruminants are relatively low (7.7% from bovine; 5.5% from sheep), and the glycosidic chains are relatively short. Methylation analysis indicates that unlike MWEL, the predominant side chain on rumen LCC is a single glucose residue bound glycosidically to the residual polyphenol. It is unclear how such a structure might arise. Other branched chain structures are also observed, some of which are linked into the polyphenol at more than one location (Fig. 7). As in the case of MWEL, oligoxylosyl residues are also present, some of which are substituted with L-arabinose. Because of the anaerobic conditions that exist in the rumen, relatively little degradation of lignin is believed to occur.

The influence of LCCs on ruminant digestion was studied by examining the solubilization of LCCs using cell-free hemicellulase complexes from the rumen (Brice & Morrison 1982). LCCs from grasses of increasing maturity were isolated and treated with cell-free rumen hemicellulases. As the lignin content increased, the extent of degradation declined, indicating that the lignin content of the LC was the overriding factor in determining its digestibility.

Degradation of lignin does not seem to occur in the anaerobic environment of the rumen, even though substantial solubilization takes place. This observation was recently confirmed using an artificial rumen reactor (Kivaisi et al. 1990). The supernatant solutions of effluents contained lignin-derived compounds that were released by rumen microorganisms.

Solubilization of LCC by microbial activity

In recent years, studies on the solubilization of lignin from grasses or wood labeled with ^{14}C phenylalanine have proliferated (Crawford 1978; Reid et al. 1982; McCarthy et al. 1984). The bulk of the radioactive label is incorporated into the lignin rather than into carbohydrate or protein of the plant, but it is clear that lignin purified from the labeled plant tissue contains significant amounts of carbohydrate. This material is probably closer to the structure of native lignin than is synthetic lignin prepared by in vitro dehydrogenative polymerization of coniferyl alcohol (Kirk et al. 1975).

Commonly, biodegradation of the MWL is followed by trapping the $^{14}CO_2$ respired from active cultures. Between 25% and 40% of the ^{14}C added to active cultures can be recovered as $^{14}CO_2$. Given that a fully aerobic organism will respire about half the carbon provided to it while incorporating the other half as cellular material, this represents the metabolism of 50% to 80% of the total lignin present. However, not all the ^{14}C lignin added to cultures is released as $^{14}CO_2$. A significant fraction of the total lignin – as much as 30% – can be recovered from solution as a polymer (Crawford et al. 1983). This material precipitates from culture filtrates following acidification to pH 3 to 5. In this characteristic, it is similar to the LCC solubilized in the rumen of cattle (Gaillard & Richards 1975). The solubilized acid precipitable polymeric lignin (APPL) has an apparent molecular weight of ≥ 20 kDa and shows signs of partial degradation. *Phanerochaete chrysosporium* will form water-soluble products from lignin isolated from aspen (Reid et al. 1982) or wheat seedlings (McCarthy et al. 1984), as will a number of actinomycetes. Organisms reported to solubilize grass lignins include *Streptomyces viridosporus, S. baddius* (Pometto & Crawford 1986), *S. cyanus, Thermomonospora mesophila,* and *Actinomadura* sp. (Mason et al. 1988; Zimmerman & Broda 1989; Mason et al. 1990).

Many different enzymatic activities from these organisms have been reported, including activities of *endo*-glucanase, xylanase, several esterases, and an extracellular peroxidase (Ramachandra et al. 1987). The roles that these enzymes play in lignin solubilization are not yet entirely clear, but various correlations have been made between the appearance of extracellular peroxidase activity and lignin solubilization or mineralization. The streptomycete lignin peroxidase has been purified and partially characterized (Ramachandra et al. 1988). It is a heme protein with an apparent molecular weight of 17.8 kDa. As such, it is appreciably smaller than the 42 kDa lignin peroxidase or the 46 kDa manganese peroxidase described in *P. chrysosporium* and other white-rot fungi (Tien & Kirk 1983; Tien & Kirk 1984; Paszczynski et al. 1985). The streptomycete enzyme has been reported to cleave a β-aryl ether model dimer in a manner similar to that observed for *P. chrysosporium*.

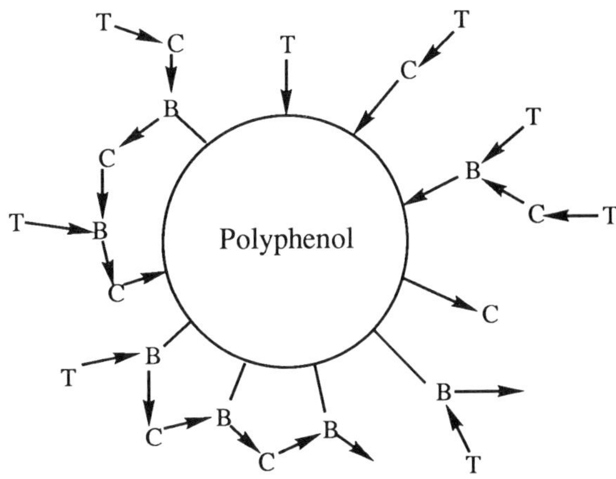

Fig. 7. Proposed structure for residual lignin carbohydrate complex from sheep rumen: T non-reducing terminal sugar residue; C chain residue; B branch point; → glycosidic linkage (arrow pointing away from C-1); — ether linkage; → free reducing group (after Conchie, Hay & Lomax 1988).

The enzyme (or enzymes) responsible for solubilizing lignin has not been fully characterized. Strains of streptomyces that produce significant quantities of APPL also produce peroxidases and cellulases (Adhi et al. 1989). With *T. mesophila*, the ability to solubilize labeled wheat lignocellulose is extracellular and inducible, but it does not correlate with xylanase or cellulase production (McCarthy et al. 1986).

Zimmerman and Broda (1989) recently reported the solubilization of lignocellulose from undigested ball-milled barley by the extracellular broth from *S. cyanus, T. mesophila,* and *Actinomadura* sp. Ground and ball-milled barley straw samples were incubated with cultures of these organisms, and weight losses were recorded. Such solubilization might simply represent the release of low molecular weight ^{14}C-labeled lignin from the carbohydrate complex. This could occur if a portion of the lignin is released as the carbohydrate is removed. Mason et al. (1988), working in the same laboratory, described the production of extracellular proteins in the broths of these cultures. They assayed fractions from a sizing gel and found that the solubilizing activity had an apparent molecular weight of about 20,000 – similar to the peroxidase reported by Ramachandra et al. (1988). Because all the xylanases and cellulases present had an apparent molecular weight of about 45,000, the authors concluded that the solubilization activity is unlikely to be from a cellulase or xylanase. An international patent application for the use of a cell-free enzyme from *S. cyanus* for the solubilization of lignocellulose has been filed. (Broda et al. 1987).

Esterases from *S. viridosporus* were reported to release *p*-coumaric and vanillic acid into the medium when concentrated, extracellular enzyme was placed on appropriately labeled substrates (Donnelly & Crawford 1988). At least eight different esterases were described. When grown on lignocellulose from corn stover, *S. viridosporus* T7A and *S. badius* 252 produced endoglucanase, xylanase, and lignin peroxidase activity (Adhi et al. 1989). Since the *p*-coumaric acid is known to be esterified essentially only to the lignin, these organisms apparently attack that substrate. In the process, they produce an APPL that can be detected in supernatant solutions.

In summary, research on the solubilization of LCC by microbial activity has periodically shown that cellulases, hemicellulases, esterases, and perhaps peroxidases all correlate with lignin solubilization. The mineralization rates and extents reported for streptomyces are relatively low, and the solubilized lignin is not extensively modified. Lignin mineralization and solubilization could, therefore be attributable to two (or more) different enzymes. These studies require more rigorous clarification.

Accessory enzymes for hemicellulose utilization

Substantial attention has been given to the principal enzymes involved in lignocellulose utilization, particularly cellulases, xylanases, and, more recently, peroxidases. A number of enzymes, however, appear to be critical in the early steps of hemicellulose utilization. These include acetyl xylan esterases, ferulic and *p*-coumaric esterases, α-L-arabinofuranosidases, and α-4-*O*-methyl glucuranosidases.

Acetyl xylan esterase was first described by Biely et al. (1985) in several species of fungi known to degrade lignocellulose and most especially in *A. pullulans*. Acetyl xylan esterase was subsequently described in a number of different microbes including *Schizophylum commune* (MacKenzie & Bilous 1988), *Aspergillus niger* and *Trichoderma reesei* (Biely et al. 1985; Poutanen & Sundberg 1988), *Rhodotorula mucilaginosa* (Lee et al. 1987) and *Fibrobacter succinogenes* (McDermid et al. 1990), and various anaerobic fungi (Borneman et al. 1990). Acetyl xylan esterase acts in a cooperative manner with endoxylanase to degrade xylans (Biely et al. 1986). This enzyme is not involved in breaking LC bonds since the acetyl esters are terminal groups.

Many different esterase activities have been described, but it has not always been apparent that the assay employed was specific for physiologically important activities. Particular care must be taken in using 4-nitrophenyl acetate as an analog of acetylated xylan since activities against 4-nitrophenyl acetate and acetylated xylan may show no correlation (Khan et al. 1990).

Relatively little is known about enzymes that are capable of releasing aromatic acids from hemicellulose. The substrates are often poorly defined, and most enzymes have been obtained only in crude preparations. Ferulic and *p*-coumaric acid esterases have been identified in extracellular broths of *S. viridosporus* (Deobald & Crawford 1987), but the activities have not always been specific (Donnelly & Crawford 1988). MacKenzie et al. (1987) first assayed for ferulic acid esterase from *Streptomyces flavogrieseus* using native substrate. In this assay, the ferulic acid was identified by HPLC. Ferulic acid esterase is produced along with α-L-arabinofuranosidase and α-4-*O*-methylglucuronidase by cells growing on xylan-containing media. With *S. flavogriseus*, oat spelts xylan was a much stronger inducer for these enzymes than cellulose. However, not all organisms respond in this manner. *Schizophyllum commune*, for example, produced more xylanase and acetyl xylan esterase when grown on avicel cellulose than when grown on xylan (MacKenzie & Bilous 1988).

The ferulic acid esterase of *S. commune* exhibits specificity for its substrate, and it has been separated from other enzymes. Borneman et al. (1990) assayed feruloyl and *p*-coumaryl esterase activities from culture filtrates of anaerobic fungi using dried cell walls of Bermuda grass (*Cyndon dactylon* [L] Pers) as a substrate. The enzyme preparations released ferulic acid more readily than they released *p*-coumaric acid from plant cell walls. Assays using methyl ferulate or methyl *p*-coumarate as substrates in place of dried cell walls showed the presence of about five times as much enzyme activity. McDermid et al. (1990) employed ethyl esters of *p*-coumarate and ferulate as substrates for these activities.

The use of a realistic model substrate can greatly facilitate purification and kinetic studies. Recently, Hatfield, Helm & Rolph (pers. comm.) synthesized 5-*O*-trans-feruloyl-α-L-arabinofuranoside in gram quantities for use as an enzyme substrate. Progress of the reaction can be followed by HPLC or TLC. The substrate is soluble in water and can be readily used for kinetic studies. Much more work needs to be done in this area, particularly with the synthesis of lignin-hemicellulose esters as model substrates.

α-L-Arabinofuranosidase catalyzes the hydrolysis of nonreducing terminal α-L-arabinofuranoside linkages in L-arabinan or from D-xylan. This enzyme has been recognized for a number of years, and the characteristics of α-L-arabinosidases of microbial and plant origin have been reviewed (Kaji 1984). Since these enzymes act primarily on terminal arabinose side chains, they are probably not directly involved in removing carbohydrate from lignin. The α-(1→ 3) linkage is hydrolyzed relatively easily. An enzyme capable of acting on L-arabinofuran substituted at *O*-5 would be of potential interest.

The α-(1→2)-4-*O*-methylglucuronic acid sidechain resists hydrolysis by xylanases (Timell, 1962), and for that matter by acid. In fact, the presence of the 4-*O*-methylglucuronic acid group stabilizes nearby xylosidic bonds in the xylan main chain, and the substituent must be removed for xylan hydrolysis to proceed. Puls et al. (1987) described a-glucuranosidases from two basidiomycetes, *Agaricus bisporus* and *Pleurotus ostreatus*; the enzyme from *A. bisporus* was partially characterized. This enzyme is evidently not one of the first to attack the xylan polymer because it is relatively large (≈ 450 kDa). Activity was optimal at pH 3.3 and 52° C. α-Glucuronidase acts in synergism with xylanases and β-xylosidases to hydrolyze glucuronoxylan. The yield of xylose greatly increases in the presence of this enzyme.

Most filamentous fungi appear to be poor producers of α-methylglucuronidases. Ishihara and Shimizu (1988) systematically screened α-glucuronidase-producing fungi in order to identify other enzymes. Of nine *Trichoderma* and five basidiomycete species (including a strain of *A. bisporus*), *Tyromyces palustris* was the best producer. Concentrated protein precipitates from cell broths were screened against a model substrate of 2-*O*-(4-*O*-methyl-α-D-glucuronopyranose)-D-xylitol. This enabled the detection of reducing group production against a very low background. Even though *T. palustris* produced more α-glucuronidase than did other fungi, total activity amounted to < 0.1 unit/ml. Moreover, the activity was very labile, even in frozen storage. *Streptomyces flavogriesus* and *S. olivochromogenes* also formed α-*O*-methylglucuronidase at low titers (MacKenzie et al. 1987; Johnson et al. 1988).

Role of glycosides in lignin degradation

Several recent reports indicate that glycosides might be involved in lignin biodegradation. Kondo & Imamura (1987) first reported that when vanilyl alcohol or veratryl alcohol were included in glucose- or cellobiose-containing media that had been inoculated with wood-rotting fungi, lignin glucosides were formed in the cellulose medium during the early phases of cultivation. Such glucosides could also be formed using a commercial β-glucosidase in place of the culture broth. β-Glucosidase biosynthesized six monomeric glucosides from lignin model compounds and cellobiose (Kondo et al. 1988). These compounds are linked through position *O*-1 of the sugar moiety to alcoholic (but not phenolic) hydroxyls of the lignin model.

In another study by Kondo & Imamura (1989a), three lignin model compounds, 4-*O*-ethylsyringylglycerol-β-syringyl ether, veratryl alcohol, and veratraldehyde, were degraded by *P. chrysosporium* and *Coriolus versicolor* in media containing either monosaccharides or polysaccharides as carbon sources. The authors found that the rate of consumption of the lignin models was much faster in polysaccharide than monosaccharide media. In media containing xylan or holocellulose, veratryl alcohol was transformed predominantly into veratryl-*O*-β-D-xyloside, which then disappeared rapidly from the medium. Veratraldehyde was first reduced to veratryl alcohol, then glycosylated, and finally consumed.

That carbohydrate is essential for the mineralization of lignin models (Kirk et al. 1976) and that the expression of lignin biodegradation is regulated by carbon catabolite repression (Jeffries et al. 1981) have been known for a long time. However, the role of carbohydrate in the assimilation of lignin degradation products has been demonstrated only recently.

One of the first enzymes implicated in lignin biodegradation was cellobiose : quinone oxidoreductase (CBQase). This enzyme catalyzes the reduction of a quinone and the simultaneous oxidation of cellobiose. Westermark & Ericksson (1974a,b) discovered this enzyme and proposed that its role might be to prevent repolymerization of lignin during degradation. More recent studies (Odier et al. 1987) have not borne this out, but the enzyme may be important nonetheless. The CBQase of *P. chrysosporium* binds very tightly to microcrystalline cellulose, but such binding does not block its ability to oxidize cellobiose, indicating that the binding and catalytic sites are in two different domains (Renganathan et al. 1990).

An essential feature of lignin biodegradation is that degradation products resulting from the activity of extracellular enzymes must be taken up by the mycelium; glycosylation by β-glucosidase seems to be an important part of this process. Whether or not sugars attached to lignin in the native substrate by nonglycosidic linkages play a similar role has not been addressed.

Glycosylation could also serve to detoxify lignin degradation products. Veratryl alcohol and vanilyl alcohol, for example, are toxic to the growth of *C. versicolor* and *T. palustris*, whereas the toxicity of the glycosides of these compounds is greatly reduced (Kondo & Imamura 1989b). The presence of a glycosyl group has also been shown to prevent polymerization of vanillyl alcohol by phenol oxidase.

Kondo et al. (1990) recently showed that glycosides can facilitate the depolymerization of dehydrogenative polymerisate (DHP) by lignin peroxidase of *P. chrysosporium* and reduce repolymerization by laccase III of *C. versicolor*. To clarify the role of glycosylation in lignin degradation, DHP and DHP-glucosides were treated with horseradish peroxidase, commercial laccase, laccase III of *C. versicolor*, and lignin peroxidase of *P. chrysosporium*. During oxidation, DHP changed color and precipitated whereas DHP glucoside only changed color. Moreover, molecular weight distribution studies showed that oxidation of DHP by enzyme preparations containing laccase or peroxidase resulted in polymerization rather than depolymerization. In contrast, enzymatically catalyzed depolymerization of DHP glucoside was observed. DHP-glucoside was also depolymerized more extensively by peroxidase than by laccase.

Enzymatic treatments of pulps

For many applications, residual lignin in kraft pulp must be removed by bleaching. Successive chlorination and alkali extraction remove the remaining lignin to leave a bright, strong pulp suitable for printing papers and other consumer products. Although chlorine bleaching solves the immediate problem of residual lignin, the chlorinated aromatic hydrocarbons produced in the bleaching step are recalcitrant and toxic. These chlorinated products are hard to remove from waste streams and trace quantities are left in the paper, so other bleaching processes have been devised. One approach is to use hemicellulases to facilitate bleaching.

Several different research groups have found that the bleaching of hardwood and softwood kraft pulp can be enhanced by xylanase. The xylanase treatment reduces chemical consumption and kappa number and increases brightness. Pine kraft pulp was delignified $>50\%$ following hemicellulase treatment and oxygen bleaching (Kantelinen et al. 1988). Small amounts of lignin were released by the enzyme treatment alone. Fungal xylanase from *Sporotricum dimorphosum* lowered the lignin content of unbleached softwood and hardwood kraft pulps (Chauvet et al. 1987). Hemicellulases having different specificities for substrate DP and side groups have been used, but more data are needed on the effects of mannanases, cellulases, and other enzymes. Enzyme-treated paper sheets show slight decreases in interfiber bonding strength. The mechanical strength of fibers is not affected, but interfiber bonding decreases if cellulases are present. This has been confirmed with the pretreatment of pulp with a xylanase from the thermophilic actinomycete *Saccharomonospora viridis* (Roberts et al. 1990). Viscosity decreases with some enzyme treatments, but only limited hemicellulose hydrolysis is necessary to enhance lignin removal. In the absence of cellulase, xylanase treatment increases viscosity. Pulps treated with cloned xylanase from *Bacillus subtilis* have retained viscosity and strength properties while lignin removal has been facilitated (Jurasek & Paice 1988; Paice et al. 1988a,b).

Conclusions

Covalent lignin-carbohydrate linkages can be ester bonds through the free carboxy of uronic and aromatic acids or ether linkages through sugar hydroxyls. Sugar hydroxyls include the primary hydroxyl of L-arabinose (*O*-5) or of D-glucose or D-mannose (*O*-6) or the secondary hydroxyl, as in the case of the *O*-2 or *O*-3 of xylose. Some linkages through the glycosidic hydroxyl (*O*-1) also appear to exist. Ester linkages can occur through the carboxyl group of uronic, ferulic, *p*-coumaric or *p*-hydroxybutyric acids. In all cases, the α-carbon of the phenylpropane subunit in lignin appears to be involved in native lignin-carbohydrate bonds, but further study is needed. Many ester linkages are disrupted by mild alkali, but a significant number of alkali-stable bonds are present. Although some microbial esterases and other enzymes have been shown to attack and solubilize lignin from lignocellulose, their substrate specificities have not been fully characterized. The positive effect of glycosylation on assimilation and degradation of lignin model compounds is supported by long-standing evidence that carbohydrate is necessary for lignin mineralization. The role of carbohydrates linked by non-glycosidic bonds remains to be clarified. It is not clear what role, if any, uronic and aromatic acid esters might play in facilitating or hindering lignin biodegradation. To date, no enzyme has been shown to cleave any bond between polymeric lignin and carbohydrate.

Acknowledgements

The author thanks John Obst of the USDAFS Forest Products Laboratory, Ron Hatfield of the USDA Dairy Forage Research Center, and John Ralph of the University of Wisconsin, Madison, for useful discussions and for critical comments on the manuscript.

References

Adhi TP, Korus RA & Crawford DL (1989) Production of major extracellular enzymes during lignocellulose degradation by two *Streptomyces* in agitated submerged culture. Appl. Environ. Microbiol. 55: 1165–1168

Akin DE, & Rigsby LL (1985) Influence of phenolic acids on rumen fungi. Agronomy J. 77: 180–182

Atushi K, Azuma J-I & Koshijima T (1984) Lignin-carbohydrate complexes and phenolic acids in bagasse. Holzforschung 38: 141–149

Azuma J-I & Koshijima T (1988) Lignin-carbohydrate complexes from various sources. Methods Enzymology 161:12–18

Azuma J-I, Takahashi N & Koshijima T (1981) Isolation and characterization of lignin-carbohydrate complexes from the milled-wood lignin fraction of *Pinus densiflora* Sieb et Zucc. Carbohyd. Res. 93: 91–104

Biely P, Krátky Z & Vrsanská M (1981) Substrate-binding site of endo-1,4-β-xylanase of the yeast *Cryptococcus albidus*. Eur. J. Biochem. 119: 559–564

Biely P, Puls J & Schneider H (1985) Acetyl xylan esterases in fungal xylanolytic systems. FEBS 186: 80–84

Biely P, MacKenzie CR, Puls J & Schneider H (1986) Cooperativity of esterases and xylanases in the enzymatic degradation of acetyl xylan. Bio/Technology 4: 731–733

Borneman WS, Hartley RD, Morrison WH, Akin DE & Ljungdahl LG (1990) Feruloyl and *p*-coumaroyl esterase from anaerobic fungi in relation to plant cell wall degradation. Appl. Microbiol. Biotechnol. 33: 345–351

Brice RE & Morrison IM (1982) The degradation of isolated hemicelluloses and lignin-hemicellulose complexes by cell-free rumen hemicellulases. Carbohyd. Res. 101: 93–100

Broda PMA, Mason JC & Zimmerman WK (1987) Decomposition of lignocellulose. International Patent WO 87/06609

Chauvet J-M, Comtat J & Noe P (1987) Assistance in bleaching of never-dried pulps by the use of xylanases: Consequences on pulp properties. 4th Intl. Symp. Wood Pulping Chem. (Paris), Poster Presentations Vol 2: 325–327

Chesson A (1988) Lignin-polysaccharide complexes of the plant cell wall and their effect on microbial degradation in the rumen. Animal Feed Sci. Technol. 21: 219–228

Chesson A, Gordon AH & Lomax JA (1983) Substituent groups linked by alkali-labile bonds to arabinose and xylose residues of legume, grass and cereal straw cell walls and their fate during digestion by rumen microorganisms. J. Sci. Food. Agric. 34: 1330–1340

Comtat J, Joseleau J-P, Bosso C & Barnoud (1974) Characterization of structurally similar neutral and acidic tetrasaccharides obtained from the enzymic hydrolysate of a 4-*O*-methyl-D-glucurono-D-xylan. Carbohyd. Res. 38:217–224.

Conchie J, Hay AJ & Lomax JA (1988) Soluble lignin-carbohydrate complexes from sheep rumen fluid: Their composition and structural features. Carbohyd. Res. 177: 127–151

Crawford D (1978) Lignocellulose decomposition by selected *Streptomyces* strains. Appl. Environ. Microbiol. 35: 1041–1045

Crawford DL, Pometto AL & Crawford RL (1983) Lignin degradation by *Streptomyces viridosporus*: Isolation and characterization of a new polymeric lignin degradation intermediate. Appl. Environ. Microbiol. 45: 898–904

Donnelly PK & Crawford DL (1988) Production by *Streptomyces viridosporus* T7A of an enzyme which cleaves aromatic acids from lignocellulose. Appl. Environ. Microbiol. 54: 2237–2244

Deobald LE & Crawford DL (1987) Activities of cellulase and other extracellular enzymes during lignin solubilization by *Streptomyces viridosporus*. Appl. Microbiol. Biotechnol. 26: 158–163

Ericksson Ö, Goring DAI & Lindgren BO (1980) Structural studies on the chemical bonds between lignins and carbohydrates in spruce wood. Wood Sci. Technol. 14: 267–279.

Ford CW (1989) A feruloylated arabinoxylan liberated from cell walls of *Digitaria decumbens* (pangola grass) by treatment with borohydride. Carbohyd. Res. 190: 137–144

Gaillard BDE & Richards GN (1975) Presence of soluble lignin-carbohydrate complexes in the bovine rumen. Carbohyd. Res. 42: 135–145

Higuchi T, Ioto Y, Shimada M & Kawamura I (1967) Chemical properties of milled wood lignin of grasses. Phytochemistry 6: 1551–1556

Ishihara M & Shimizu K (1988) α-(1→ 2)-glucuronidase in the enzymatic saccharification of hardwood xylan I. Screening of α-glucuronidase producing fungi. Mokuzai Gakkaishi 34: 58–64

Ishii T & Hiroi T (1990) Isolation and characterization of feruloylated arabinoxylan oligosaccharides from bamboo shoot cell-walls. Carbohyd. Res. 196: 175–183

Iversen T, Westermark U & Samuelsson B (1987) Some comments on the isolation of galactose-containing lignin-carbohydrate complexes. Holzforschung 41: 119–121

Jeffries TW, Choi S & Kirk TK (1981) Nutritional regulation of lignin degradation by *Phanerochaete chrysosporium*. Appl. Environ. Microbiol. 42: 290–296

Johnson KG, Harrison BA, Schneider H, MacKenzie CR & Fontana JD (1988) Xylan-hydrolyzing enzymes from *Streptomyces* spp. Enzyme Microb. Technol. 10: 403–409

Joseleau J-P, & Gancet C (1981) Selective degradations of the lignin-carbohydrate complex from aspen wood. Svensk Papperstidning 84: R123–R127

Jurasek L & Paice MG (1988) Biological beaching of pulp (pp 11–13). Tappi Internat. Pulp Bleach. Conf., Orlando, Florida

Kaji A (1984) L-Arabinosidases. Advan. Carbohyd. Chem. Biochem. 42: 383–394

Kantelinen A, Rättö M, Sundquist J, Ranua M, Viikari L & Linko M (1988) Hemicellulases and their potential role in bleaching (pp 1–9). Tappi Internat. Pulp Bleaching Conf., Orlando, Florida

Kato A, Azuma JI & Koshijima T (1987) Isolation and identification of a new feruloylated tetrasaccharide from bagasse lignin-carbohydrate complex containing phenolic acid. Agric. Biol. Chem. 51: 1691–1693

Khan AW, Lanm KA & Overend RP (1990) Comparison of natural hemicellulose and chemically acetylated xylan as substrates for the determination of acetyl-xylan esterase activity in *Aspergilli*. Enzyme Microb. Technol. 12: 127–131

Kirk TK, Connors WJ, Bleam RD, Hackett WF & Zeikus JG (1975) Preparation and microbial decomposition of synthetic [^{14}C] lignins. Proc. Natl. Acad. Sci. U.S.A. 72: 2515–2519

Kirk TK, Connors WJ & Zeikus JG (1976) Requirement for a growth substrate during lignin decomposition by two wood-rotting fungi. Appl. Environ. Microbiol. 32: 192–194

Kivaisi AK, Op den Camp HJM, Lubberding HJ, Boon JJ & Vogels GD (1990) Generation of soluble lignin-derived compounds during degradation of barley straw in an artificial rumen reactor. Appl. Microbiol. Biotechnol. 33: 93–98

Kondo R & Imamura H (1987) The formation of model glycosides by wood-rotting fungi. Lignin enzymatic and microbial degradation. INRA, Paris

Kondo R & Imamura H (1989a) Formation of lignin model xyloside in polysaccharides media by wood-rotting fungi. Mokuzai Gakkaishi 35: 1001–1007

Kondo R & Imamura H (1989b) Model study on the role of the formation of glycosides in the degradation of lignin by wood-rotting fungi. Mokuzai Gakkaishi 35: 1008–1013

Kondo R, Imori T & Imamura H (1988) Enzymatic synthesis of glucosides of monomeric lignin compounds with commercial β-glucosidase. Mokuzai Gakkaishi 34: 724–731

Kondo R, Imori T, Imamura H & Kishida T (1990) Polymerization of DHP and depolymerization of DHP glucoside by lignin oxidizing enzymes. J. Biotechnol. 13: 181–188

Koshijima T, Watanabe T & Yaku T (1989) Structure and properties of the lignin-carbohydrate complex polymer as an amphipathic substance. In: Glasser WG & Sarkanen S (Eds) Lignin Properties and Materials. ACS Symposium Ser. 397 (pp 11–28). American Chemical Society, Washington, D.C.

Lee H, To RJB, Latta RK, Biely P & Schneider H (1987) Some properties of extracellular acetylxylan esterase produced by the yeast *Rhodotorula mucilaginosa*. Appl. Environ. Microbiol. 53: 2831–-2834

MacKenzie CR & Bilous D (1988) Ferulic acid esterase activity from *Schizophyllum commune*. Appl. Environ. Microbiol. 54: 1170–1173

MacKenzie CR, Bilous D, Schneider H & Johnson KG (1987) Induction of cellulolytic and xylanolytic enzyme systems in *Streptomyces* spp. Appl. Environ. Microbiol. 53: 2835–2839

Markwalder HU & Neukom H (1976) Diferulic acid as a possible crosslink in hemicelluloses from wheat germ. Phytochemistry 15: 836–837

Mason JC, Richards M, Zimmerman W & Broda P (1988) Identification of extracellular proteins from actinomycetes responsible for the solubilization of lignocellulose. Appl. Microbiol. Biotechnol. 28: 276–280

Mason JC, Birch OM & Broda P (1990) Preparation of ^{14}C-radiolabelled lignocelluloses from spring barley of differing maturities and their solubilization by *Phanerochaete chrysosporium* and *Streptomyces cyanus*. J. Gen. Microbiol. 136: 227–232

McCarthy AJ, MacDonald MJ, Paterson A & Broda P (1984) Degradation of [^{14}C] lignin-labelled wheat lignocellulose by white-rot fungi. J. Gen. Microbiol. 130: 1023–1030

McCarthy AJ, Paterson A & Broda P (1986) Lignin solubilization by *Thermonospora mesophila*. Appl. Microbiol. Biotechnol. 24: 347–352

McDermid KP, MacKenzie CR & Forsberg CW (1990) Esterase activities of *Fibrobacter succinogenes* subsp *Succinogenes* S85. Appl. Environ. Microbiol. 56: 127–132

Meshitsuka G, Lee ZZ, Nakano J & Eda S (1983) Contribution of pectic substances to lignin-carbohydrate bonding. Int. Symp. Wood Pulping Chem. 1: 149–152

Minor JL (1982) Chemical linkage of pine polysaccharides to lignin. J. Wood Chem. Technol. 2(1): 1–16

Morison IM (1974) Structural investigation on the lignin-carbohydrate complexes of *Lolium perene*. Biochem J. 139: 197–204

Mukoyoshi SI, Azuma JI and Koshijima T (1981) Lignin-carbohydrate complexes from compression wood of *Pinus densiflora* Sieb et. Zucc. Holzforschung 35: 233–240

Neilson MJ & Richards GN (1982) Chemical structures in a lignin-carbohydrate complex isolated from bovine rumen. Carbohyd. Chem. 104: 121–138

Obst JR (1982) Frequency and alkali resistance of lignin-carbohydrate bonds in wood. Tappi 65(4): 109–112

Odier E, Mozuch M, Kalyanaraman B & Kirk TK (1987) Cellobiose: quinone oxidoreductase does not prevent oxidative coupling of phenols or polymerization of lignin by ligninase. Les Colloques de l'INRA, No. 40. Dekker 131–136

Paice MG, Bernier R & Jurasek L (1988a) Viscosity-enhancing bleaching of hardwood kraft pulp with xylanase from a cloned gene. Biotechnol. Bioeng. 32: 235–239

Paice MG, Bernier R & Jurasek L (1988b) Bleaching hardwood kraft with enzymes from cloned systems. CPPA Ann. Mtg. (Montreal) preprints 74A: 133–136

Paszczynski A, Huynh V-B & Crawford R (1985) Enzymatic activities of an extracellular manganese-dependent peroxidase from *Phanerochaete chrysosporium*. FEMS Microbiol. Lett. 29: 37–41

Pometto AL & Crawford DL (1986) Catabolic fate of *Streptomyces viridosporus* T7A-produced, acid-precipitable polymeric lignin upon incubation with ligninolytic *Streptomyces* species and *Phanerochaete chrysosporium*. Appl. Environ. Microbiol. 51: 171–179

Poutanen K & Sundberg M (1988) An acetyl esterase of *Trichoderma reesei* and its role in the hydrolysis of acetyl xylans. Appl. Microbiol. Biotechnol. 28: 419–424

Puls J, Schmidt O & Granzow C (1987) α-Glucuronidase in two microbial xylanolytic systems. Enzyme Microb. Technol. 9: 83–88

Ramachandra M, Crawford DL & Pometto AL (1987) Extracellular enzyme activities during lignocellulose degradation by *Streptomyces* spp.: A comparative study of wild-type and genetically manipulated strains. Appl. Environ. Microbiol. 53: 2754–2760

Ramachandra M, Crawford DL & Hertel G (1988) Characterization of an extracellular lignin peroxidase of the lignocellulolytic actinomycete *Streptomyces viridosporus*. Appl. Environ. Microbiol. 54: 3057–3063

Reid ID, Abrams GD & Pepper JM (1982) Water soluble products from the degradation of aspen lignin by *Phanerochaete chrysosporium* Can. J. Bot. 60: 2357–2364

Renganathan V, Usha SN, & Lindenburg F (1990) Cellobiose-oxidizing enzymes from the lignocellulose-degrading basidiomycete *Phanerochaete chrysosporium*: Interaction with microcrystalline cellulose. Appl. Microbiol. Biotechnol. 32: 609–613

Roberts JC, McCarthy AJ, Flynn NJ & Broda P (1990) Modification of paper properties by the pretreatment of pulp with *Saccharomonospora viridis* xylanase. Enzyme Microb. Technol. 12: 210–213

Scalbert A, Monties B, Lallemand JY, Guittet E & Rolando C (1985) Ether linkage between phenolic acids and lignin fractions from wheat straw. Phytochemistry 24: 1359–1362

Shimada M, Fukuzuka T & Higuchi T (1971) Ester linkages of *p*-coumaric acid in bamboo and grass lignins. Tappi 54: 72–78

Smith DCC (1955) Ester groups in lignin. Nature 176: 267–268

Takahaski N & Koshijima T (1988) Ester linkages between lignin and glucuronoxylan in a lignin-carbohydrate complex from beech (*Fagus crenata*) wood. Wood Sci. Technol. 22: 231–241

Tanabe H & Kobayashi Y (1986) Enzymatic maceration mechanism in biochemical pulping of mitsumata (*Edgeworthia papyrifera* Sieb. et Zucc.) bast. Agric. Biol. Chem. 50: 2779–2784

Tanabe H & Kobayashi Y (1987) Effect of lignin-carbohydrate complex on maceration of mitsumata (*Edgeworthia papyrifera* Sieb. et Zucc.) bast by pectinolytic enzymes from *Erwinia carotovora*. Holzforschung 41: 395–399

Tanabe H & Kobayashi Y (1988) Aggregate of pectic substances and lignin-carbohydrate complex in mitsumata (*Edgeworthia papyrifera* Sieb. et Zucc.) bast and its degradation by pectinolytic enzymes from *Erwinia cartovora*. Holzforschung 42: 47–52

Tien M & Kirk TK (1983) Lignin-degrading enzyme from hymenomycete *Phanerochaete chrysosporium* Burds. Science 221: 661–663

Tien M & Kirk TK (1984) Lignin-degrading enzyme from *Phanerochaete chrysosporium*: Purification, characterization, and catalytic properties of a unique H_2O_2-requiring oxygenase. Proc. Natl. Acad. Sci. U.S.A. 81: 2280–2284

Timell TE (1962) Enzymatic hydrolysis of a 4-*O*-methylglucuronoxylan from the wood of white birch. Holzforschung 11: 436–447

Wang PY, Bolker HI & Purves CB (1967) Uronic acid ester groups in some softwoods and hardwoods. Tappi 50(3): 123–124

Watanabe T & Koshijima T (1988) Evidence for an ester linkage between lignin and glucuronic acid in lignin-carbohydrate complexes by DDQ-oxidation. Agric. Biol. Chem. 52: 2953–2955

Watanabe TJ, Ohnishi Y, Kaizu YS & Koshijima T (1989) Binding site analysis of the ether linkages between lignin and hemicelluloses in lignin-carbohydrate complexes by DDQ-oxidation. Agric. Biol. Chem. 53: 2233–2252

Westermark U & Ericksson KE (1974a) Carbohydrate-dependent enzymic quinone reduction during lignin degradation. Acta Chem. Scand. B 28: 204–208

Westermark U & Ericksson KE (1974b) Cellobiose-quinone oxidoreductase, a new wood-degrading enzyme from white-rot fungi. Acta Chem. Scand. B 28: 209–214

Zimmerman W & Broda P (1989) Utilization of lignocellulose from barley straw by actinomycetes. Appl. Microbiol. Biotechnol. 30: 103–109

Biodegradation 1: 177–190, 1990.

Physiology of microbial degradation of chitin and chitosan

Graham W. Gooday
Department of Molecular and Cell Biology, University of Aberdeen, Marischal College, Aberdeen, AB9 1AS, UK

Key words: biodegradation, chitin, chitin deactylase, chitinase, chitosan, chitosanase

Abstract

Chitin is produced in enormous quantities in the biosphere, chiefly as the major structural component of most fungi and invertebrates. Its degradation is chiefly by bacteria and fungi, by chitinolysis via chitinases, but also via deacetylation to chitosan, which is hydrolysed by chitosanases. Chitinases and chitosanases have a range of roles in the organisms producing them: autolytic, morphogenetic or nutritional. There are increasing examples of their roles in pathogenesis and symbiosis. A range of chitinase genes have been cloned, and the potential use for genetically manipulated organisms over-producing chitinases is being investigated. Chitinases also have a range of uses in processing chitinous material and producing defined oligosaccharides.

Introduction: chitin and chitosan

Chitin, the (1–4)-β-linked homopolymer of *N*-acetyl-D-glucosamine (Fig. 1), is produced in enormous amounts in the biosphere. A recent working estimate for both annual production and steady-state amount is of the order of 10^{10} to 10^{11} tons (Gooday 1990a). Chitin is utilized as a structural component by most species alive today. Its phylogenetic distribution is clearly defined:

Prokaryotes. Despite its chemical similarity to the polysaccharide backbone of peptidoglycan, chitin has only been reported as a possible component of streptomycete spores and the stalks of some prosthecate bacteria.

Protista. Chitin provides the tough structural material for many protists: in cyst walls of some ciliates and amoebae; in the lorica walls of some ciliates and chrysophyte algae: in the flotation spines of centric diatoms; and in the walls of some chlorophyte algae and oomycete fungi (Gooday 1990a).

Fungi. Chitin appears to be ubiquitous in the fungi (Bartnick-Garcia & Lippman 1982). Reported exceptions, such as *Schizosaccharomyces*, prove to have small but essential amounts of chitin (H. Sietsma, pers. comm.). *Pneumocystis carinii*, of uncertain affinity, has chitin the walls of its cysts and trophozoites (Walker et al. 1990).

Animals. Chitin is the characteristic tough material playing a range of structural roles among most invertebrates (Jeuniaux 1963, 1982). It is absent from vertebrates.

Plants. Chitin sensu stricto is probably absent from plants, but polymers rich in (1–4)-β-linked *N*-acetylglucosamine have been reported (Benhamou & Asselin 1989).

Chitin occurs in a wide variety of types. Three hydrogen-bonded crystalline forms have been characterized: α-chitin with antiparallel chains, β-chitin with parallel chains and γ-chitin with a three-chain unit cell, two 'up' – one 'down' (Blackwell

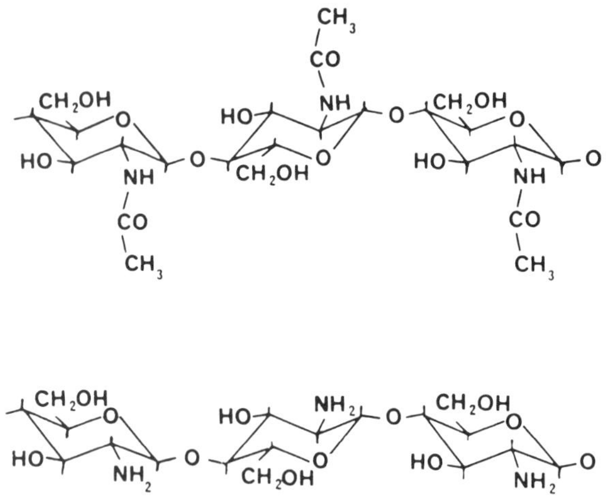

Fig. 1. Structures of chitin (top) and chitosan (bottom).

1988). α-Chitin is by far the most common, being the form found in fungi and most protistan and invertebrate exoskeletons. The importance of physical form to biological function is indicated by squid, *Loligo,* having α-chitin in its tough beak, β-chitin in its rigid pen, and γ-chitin in its flexible stomach lining.

With one exception, the chitin of diatom spines, chitin is always found cross-linked to other structural components. In fungal walls it is cross-linked covalently to other wall components notably β-glucans (Sietsma et al. 1986; Surarit et al. 1988). In insects and other invertebrates, the chitin is always associated with specific proteins, with both covalent and noncovalent bonding, to produce ordered structures (Blackwell 1988). There are often also varying degrees of mineralization, in particular calcification, and sclerotization, involving interactions with phenolic and lipid molecules (Poulicek et al. 1986; Peter et al. 1986).

Another modification of chitin is its deacetylation to chitosan, the (1–4)-β-linked polymer of D-glucosamine (Fig. 1). This is mediated by the enzyme chitin deacetylase. In the fungi this occurs in the Mucorales, where chitosan is a major component of the cell wall (Datema et al. 1977; Davis & Bartnicki-Garcia 1984) and in *Saccharomyces cerevisiae,* where it is a major component of ascospore walls. The biological significance of this deacetylation in fungi may be to give them added resistance to lysis by chitinolytic organisms. Deacetylation also occurs in arthropods, where its occurrence seems to be related to chitinous structures that undergo subsequent expansion, such as the abdominal cuticle of physogastric queen termites, and eye-lens cuticles (Aruchami et al. 1986).

Pathways of chitin degradation

The vast annual production of chitin is balanced by an equal rate of recycling. The bulk of this chitin degradation is microbial; in the sea chiefly by bacteria-free-living and in association with animal guts; in the soil chiefly by fungi and bacteria. Their biochemical pathways have been reviewed by Davis & Eveleigh (1984). Organisms that degrade chitin solely by hydrolysis of glycosidic bonds are known as chitinolytic; a more general term, not specifying the mechanism, is 'chitinoclastic'.

The best-studied pathway is the action of the chitinolytic system, of hydrolysis of the glycosidic bonds of chitin. Exochitinase cleaves diacetylchitobiose units from the non-reducing end of the polysaccharide chain. Endochitinase cleaves glycosidic linkages randomly along the chain, eventually giving diacetylchitibiose as the major product, together with some triacetylchitotriose. There may not always be a clear distribution between these two activities (see also Davis & Eveleigh 1984), as the action of these enzymes is dependant on the nature of the substrate. Thus the pure crystalline β-chitin of diatom spines is degraded only from the ends of the spines by *Streptomyces* chitinase complex, to yield only diacetylchitobiose, whereas colloidal (reprecipitated) chitin is degraded to a mixture of oligomers and diacetylchitobiose (Lindsay & Gooday 1985a). Lysozyme has a low endochitinolytic activity, but can readily be distinguished from chitinases as it readily hydrolyses *Micrococcus* peptidoglycan whereas they do not. Diacetylchitobiose (often called chitobiose, but beware confusion with the product of chitosanase) is hydrolysed to *N*-acetylglucosamine by β-*N*-acetylglucosaminidase. Some β-*N*-acetylglucosaminidases can also act weakly as exochitinases, cleaving monosaccharide units from the non-reducing ends of chitin chains.

Together, the chitinases and β-*N*-acetylglucosaminidases are known as 'the chitinolytic system'.

An alternative system for degrading chitin is via deacetylation to chitosan which is hydrolysed by chitosanase to give chitobiose, glucosaminyl-(1–4)-β-glucosaminide, which in turn is hydrolysed by glucosaminidase to glucosamine. This pathway appears to be important in some environments, for example in estuarine sediments, where chitosan is a major organic constituent (Hillman et al. 1989a, b; Gooday et al. 1991). As yet, there are no reports of a third possible pathway, involving deamination of the aminosugars (Davis & Eveleigh 1984).

Autolytic and morphogenetic chitinolysis

Where investigated in detail, all chitin-containing organisms also produce chitinolytic enzymes. In some cases, such as arthropod moulting, a role is obvious. Microbial examples include the basidiomycete fungi, the inkcaps *Coprinus* species and the puff-balls, *Lycoperdon* species, where massive autolysis follows basidiospore maturation (Iten & Matile 1970; Tracey 1955). In the case of *Coprinus*, the basidiospore discharge starts at the outermost, i.e. bottom, edges of the gills, and the gills then progressively autolyse upwards so that the spores are always released with only a fraction of a millimetre to fall into the open air for dispersal. Thus, unlike most agarics, precise vertical orientation of the gills is not required, and they are not geotropic. In the case of *Lycoperdon*, the spore-producing gleba autolyses to give a capillitium of long dry springy hyphae packed with dry spores. Raindrops cause the puff-ball to act like bellows, expelling puffs of spores into the open air. Autolytic chitinases must also act in consort with other lytic enzymes to allow plasmogamy during sexual reproduction in fungi, for example to break down the gametangial walls in the Mucorales (Sassen 1965), and to break down septa to allow nuclear migration during dikaryotization in basidiomycetes (Janszen & Wessels 1970). The accumulation of autolytic enzymes in culture filtrates of senescent fungal cultures in well-documented (Reyes et al. 1984; Isaac & Gokhale 1982) but it is unclear to what extent the chitin is recycled by these mycelia.

Chitinous fungi also produce chitinases during exponential growth. Examples include *Mucor* (Humphreys & Gooday 1984a,b,c; Gooday et al. 1986), *Neurospora crassa* (Zarain-Herzberg & Arroyo-Begovich 1983) and *Candida albicans* (Barrett-Bee & Hamilton 1984).

Possible roles for these chitinases are discussed by Gooday et al. (1986) and Gooday (1990b). They include:

Maturation of chitin microfibrils. The form of microfibrils in the wall differs in different fungi and between different life stages in the same fungus (Gow & Gooday 1983). The formation of antiparallel α-chitin microfibrils of particular orientation, length and thickness may require modelling of the chitin chains by chitinases, both by their lytic activities and their transglycosylase activities (Gooday & Gow 1991). Their transglycosylase activities may also have a role in covalently linking chains with other wall polysaccharides.

Apical growth. The 'unitary model' of hyphal growth (Bartnicki-Garcia 1973) envisages a delicate balance between wall synthesis and wall lysis, allowing new chitin chains to be continually inserted into the wall, with concomitant lysis of pre-existing chains to allow this. There is much circumstantial evidence for the role of chitinases and other lytic enzymes in this process (Gooday & Gow 1991), but as yet there is no direct evidence. The membrane-bound *Mucor* chitinase studied by Humphreys & Gooday (1984a, b, c) shared with chitin synthase the property of being activatable by trypsin, i.e. being zymogenic, suggesting that the two enzymes could be co-ordinately regulated, as would be required for orderly chitin deposition.

Branching. It is generally accepted that chitinase action will be required to form a branch. The cylindrical wall of a hypha, unlike the apex, is a rigid structure. Its chitin microfibrils are wider, more crystalline, and are cross-linked with other wall components (Wessels 1988). The site of the new branch must be weakened to allow a new apex to be formed, and lytic enzymes are obvious contenders for this process.

Spore germination. Germination of fungal spores, and indeed hatching of protozoal cysts, requires the breaching of the wall. It seems likely that chitinases have a role in this process in at least some cases, for example in sporangiospore germination of *Mucor mucedo,* where the initial spherical growth is accompanied by a co-ordinated activation of chitinase and chitin synthase (Gooday et al. 1986).

In the budding yeast *Saccharomyces cerevisiae*, chitin is mostly confined to the septum separating the bud from the mother cell, where it is a major component. Elango et al. (1982) showed that chitinase is a periplasmic enzyme in these yeast cells, and suggested that it plays a role in cell separation. More direct evidence for this is provided by the findings that the chitinase inhibitors, allosamidin and demethylallosamidin, inhibit cell separation and lytic damage during budding (Cabib et al. 1990; Sakuda et al. 1990).

Nutritional chitinolysis

Bacteria

Chitinolytic bacteria are widespread in all productive habitats. Chitinases are produced by many genera of Gram negative and Gram positive bacteria, but not by Archaebacteria (Gooday 1979; Berkeley 1979; Monreal & Reese 1969).

The sea produces vast amounts of chitin, chiefly as carapaces of zooplankton, which are regularly moulted as the animals grow. Most of this chitin is produced near to the surface, and studies have shown that its recycling occurs both in the water column and in sediments (reviewed by Gooday 1990a). The rate of degradation will be enhanced by phenomena of adherence of chitinolytic microflora and by passage through animals guts. The importance of these processes is highlighted by the repeated finding of chitinolytic bacteria, principally of the genera *Vibrio* and *Photobacterium*, associated with zooplankton and particulate matter (e.g. Hood & Meyers 1977). Estimations of population densities of chitinolytic bacteria, both as total counts and as percentages of total heterotrophs, have shown considerable variation, but consistently higher counts have been reported from marine sediments than from the overlying seawater (Gooday 1990). Studies such as that by Helmke & Weyland (1986) conclude that indigenous bacteria are capable of decomposing chitin particles throughout the depth of the Antarctic Ocean, as are chitinases produced in surface waters and transported down by sinking particles.

Estuaries are particularly productive: Reichardt et al. (1983) isolated 103 strains of chitinolytic bacteria from the estuarine upper Chesapeake Bay. Of these, 44 were yellow-orange pigmented *Cytophaga*-like bacteria , with a range of salt requirements. Others were vibrios, pseudomonads and *Chromobacterium* strains. Chan (1970) presented studies of chitinolytic bacteria from Puget Sound: genera identified, in decreasing order of abundance, were *Vibro, Photobacterium, Aeromonas, Cytophaga, Streptomyces, Photobacterium, Bacillus* and *Chromobacterium*.Pel & Gottschal (1986a, b; 1989) and Pel et al (1989, 1990) have investigated chitinolysis by *Clostridium* strains isolated from sediments and the anoxic intestine of plaice from the Eems-Dollard estuary. They found that in pure culture chitin was degraded slowly; diacetylchitobiose accumulated, but soon disappeared as *N*-acetylglucosamine accumulated. They suggested that the *Clostridium* strains are specialised utilizers of diacetylchitobiose, and accumulation of *N*-acetylglucosamine represents non-utilizable monomers appearing during random hydrolysis of chitin oligomers. Chitin degradation was greatly enhanced by coculture with other bacteria from the sediments. One aspect of this enhancement they suggest is the release of stimulatory growth factors, such as a thioredoxin-type compound that maintained the reduced state of essential sulphydryl groups in the chitinolytic system. Interspecies interactions may also play a role for this bacterium if it is exposed to oxygen in the upper layers of sediments, as accumulating mono- and disaccharides could provide substrates for facultative aerobic bacteria, which would consume oxygen to render the microenvironment anaerobic again.

Chitinolytic bacteria are also abundant in freshwaters, characteristic genera in the water column being *Serratia, Chromobacterium, Pseudomonas, Flavobacterium,* and *Bacillus*, with *Cytophaga johnsonae* and actinomycetes in sediments (Gooday 1990a).

The soil contains many chitinous animals and fungi as its normal living components, and chitinolytic bacteria can be isolated readily. The numbers and types reported vary greatly with different soils and methods of isolation, but major genera are *Pseudomonas, Aeromonas, Cytophaga johnsonae, Lysobacter, Arthrobacter, Bacillus* and actinomycetes (Gooday 1990a).

When grown in liquid culture, most of the chitinolytic bacteria secrete chitinases into the medium. *Cytophaga johnsonae*, a ubiquitous soil organism, characteristically binds to chitin as it degrades it. Wolkin & Pate (1985) describe a class on nonmotile mutants with an interesting pleiotropy. They are all unable to digest and utilize chitin, as well as being resistant to phages that infect the parental strain and having relatively non-adherent and non-hydrophobic surfaces compared with wild-type strains. The authors conclude that all characteristics associated with this pleiotropy require moving cell surfaces, and that chitin digestion requires some feature of this, presumably involving enzymatic contact between bacterium and substrate. Pel & Gottschal (1986a) illustrate direct contact between cells of the chitinolytic *Clostridium str*. 9.1 and chitin fibrils, and as for cellulolytic *Clostridium* species, this may involve specific enzymatic structures on the cell surface. Particular attention has been paid to adsorption of the pathogenic but also chitinolytic *Vibrio* species. Kaneko & Colwell (1978) describe strong adsorption to chitin of *Vibrio parahaemolyticus* from the estuarine Chesapeake Bay. They suggest that this has an ecological as well as digestive significance to the bacteria, as the adsorption was reduced by increasing values of salinity and pH from those of the estuary to those of sea-water. This phenomenon would favour retention of the bacteria within the estuary. Bassler et al. (1989) have found that not only does *Vibrio furnissii* adhere to chitin, but also it shows specific chemotaxis to chitin oligosaccharides (monosaccharide to hexasaccharide), with at least two or three distinct chemoreceptors. In contrast it shows slight to no chemotaxis to a range of other nutrient sources, such as glycerol, lactate and amino acids, with the exception of L-glutamic acid.

Where investigated, chitinase production by bacteria has been shown to be inducible by chitin oligomers and low levels of *N*-acetylglucosamine (Jeaniaux 1963; Monreal & Reese 1969; Kole & Altosaar 1985).

Fungi

Chitinolytic fungi are readily isolated from soils, where they rival or even exceed the chitinolytic activities of bacteria. Most common are Mucorales, especially *Mortierella* spp. and Deuteromycetes and Ascomycetes, especially the genera *Aspergillus, Trichoderma, Verticillium, Thielavia, Penicillium* and *Humicola* (Gooday 1990a). These fungi characteristically have inducible chitinolytic systems (Sivan & Chet 1989). Baiting of freshwater sites with chitin yields a range of chitinolytic fungi, interesting members of which are the chytrids, such as *Chytriomyces* species (Reisert & Fuller 1962), and *Karlingia astereocysta*, which has a nutritional requirement for chitin that can only be relieved by *N*-acetylglucosamine; i.e. it is an 'obligate chitinophile' (Murray & Lovett 1966). Fungi are rare in the sea, but the sea is rich in chitin, and Kohlmeyer (1972) described a range of fungi degrading the chitinous exoskeletons of hydrozoa. Only one could be identified, the ascomycete *Abyssomyces hydrozoicus*.

Slime moulds, protozoa and algae

The Myxomycetes, 'true slime moulds', are a rich source of lytic enzymes, and, for example, *Physarum polycephalum* produces a complex of extracellular chitinases (Pope & Davies 1979). Soil amoebae, *Hartmanella* and *Schizopyrenus* produce chitinases. These enzymes must participate in the digestion of chitinous food particles engulfed by the slime mould plasmodium and by the amoebae. Phagocytotic ciliates probably also have the capacity to digest chitin, and chitinase activities have been implicated in the unusual feeding strategies of *Ascophrys*, a chitinvorous ectosymbiont of shrimps (Bradbery et al. 1987), and *Grossglockneria*, which feeds by digesting a tiny hole through a fungal hypha and sucking out cytoplasm (Petz et al. 1986). The colourless heterotrophic diatom, *Nitzschia alba*, is also reported to digest chitin (A.E. Linkens, quoted by Hellebust & Lewin 1977).

Chitinolysis in pathogenesis and symbiosis

Pathogens of chitinous organisms characteristically produce chitinases. These can have two roles; they can aid in the penetration of the host; and they can provide nutrients directly in the form of amino sugars and indirectly by exposing other host material to enzymatic digestion. Examples include the oomycete *Aphanomyces astaci*, a pathogen of crayfish (Soderhall & Unestam 1975); the fungus *Paecilomyces lilacius*, a pathogen of nematode eggs (Dackman et al 1989); the entomopathogenic fungi, *Beauveria bassiana, Metarhizium anisopliae* and *Verticillium lecanii* (Smith & Grula 1983; St Leger et al. 1986); mycophilic fungi, *Cladobotryum* species and *Aphanocladium album* (G.W.Gooday unpubl.; Zhloba et al. 1980; Srivastava et al. 1985); the bacterial *Serratia* species, insect pathogens (Lysenko 1976; Flyg & Boman 1988); and a *Photobacterium* species causing exoskeleton lesions of the tanner crab (Baross et al. 1978). Other examples where chitinases may be implicated but have yet to be characterised include the ciliate protozoa feeding on fungi and shrimps, discussed in the previous section, and the baculoviruses infecting insect larvae. Infection by these viruses is mediated by a disruption of the chitinous peritrophic membrane by a viral factor of unknown action (Derksen & Granados 1988).

That the chitin of the gut peritrophic membrane is a site of attack by insect pathogenic bacteria is suggested by experiments with *Drosphila melanogaster* (Flyg & Boman 1988). Flies with mutations in two genes *cut* and *miniature* are more susceptible than the wild type to infection by *Serratia marcescens*. That the *cut* and *miniature* mutations lead to deficiencies in chitin content was demonstrated by showing that pupal shells from the mutant strains were more readily digested by *Serratia* chitinase, and especially by synergistic action of chitinase and protease, than those of other strains. Also a mutant bacterial strain, deficient in chitinase and protease, was much less pathogenic to the flies. Daust & Gunner (1979), studying bacterial pathogenesis of larvae of the gypsy moth, showed that the virulence of the chitinolytic bacterium strain 501B was synergistically enhanced by co-feeding the larvae with fermentative nonpathogenic bacteria. They explained this by the acid production by the fermentative bacteria having the effect of lowering the alkaline pH of the larval gut to a value that gave greater activity of the chitinase from 501B, leading to disruption of the peritrophic membrane. The sugar beet root maggot, however, has turned the chitinolysis by *Serratia* to its advantage by developing a symbiotic relationship with *S. liquefaciens* and *S. marcescens* (Iverson et al. 1984). These bacteria become embedded in the inner puparial surface, and aid the emergence of the adult fly by their digestion of the chitin of the puparium. The symbiotic bacteria are present in all developmental stages, including the eggs. Maternally inherited chitinolytic bacteria are also implicated in susceptibility of tsetse flies to infection with trypanosomes (Maudlin & Welburn 1988). The susceptible flies have infections of 'rickettsia-like organisms', which produce chitinase when in culture in insect cells. The resistance of refractory tsetse flies (lacking the bacterial infection) is ascribed to killing of the trypanosomes in the gut mediated by a lectin. Maudlin & Welburn suggest that the bacterial chitinolysis releases amino sugars that inhibit the lectin-trypanosome binding, and thus results in survival of the trypanosomes.

Chitinase production by the entomopathogenic fungi is inducible by chitin oligomers, *N*-acetylglucosamine and glucosamine (Smith & Gruler 1983;

St. Leger et al. 1986). St Leger et al. also report that chitosanase is co-induced with chitinase in *M. anisopliae*. In insect pathogenesis, chitinase has its importance in acting in synergism with proteases, and Bibochka & Khachatourians (1988) suggest that both activities are coordinately regulated. They show that low levels of *N*-acetylglucosamine will induce a serine protease in *Beauveria bassiana*, and suggest that an initial constitutive chitinase attack on the insect cuticle would yield *N*-acetylglucosamine, leading to the coordinate induction of chitinases and proteases.

Chitin in fungi and invertebrates composes a considerable part of the diet of many herbivorous and carnivorous animals. There can be three sources of chitinolytic enzymes in the animal's digestive system; from the animal itself, from the endogenous gut microflora, or from the ingested food (Gooday 1990a). Most work has been done with fish, where a typical marine fish gut microflora is dominated by chitinolytic strains of *Vibrio, Photobacterium* and enterobacteria. However, it is clear that the fish produce their own chitinases, which they use as food processing enzymes rather than directly nutritional enzymes. Thus the gut bacteria cannot be regarded as mutualistic symbionts with respect to chitin as the rumen symbionts are with respect to cellulose (Lindsay et al. 1984; Lindsay & Gooday 1985b; Gooday 1990a). With mammals the situation is less clear: whales have chitinolytic microflora in their stomachs, which may contribute to a rumen-type fermentation (Seki & Taga 1965; Herwig et al. 1984); Patton & Chandler (1975) describe digestion of chitin by calves and steers, implying a chitinolytic rumen flora; and Kuhl et al. (1978) found elevated caecal weights in chitin-fed rats, suggesting participation of intestinal bacteria in chitin digestion.

Among invertebrates, chitin digestion is widespread with or without participation of a microbial chitinolytic flora (Jeuniaux 1963). Borkott & Insam (1990), working with the soil springtail, *Folsomia candida*, conclude that at least in this arthropod there is a mutualistic symbiosis with its gut chitinolytic bacteria, *Xanthomonas* and *Curtobacterium* species. Thus the steady increase in biomass in animals fed every four days with chitin plus yeast extract was prevented by treatment with the antibiotic tetracycline. In a food preference experiment, the animals chose to feed on chitin agar strips that had been pre-inoculated with the chitinolytic bacteria or the animal's faeces, suggesting that some pre-digestion of the chitin was aiding its utilization by the animal.

Degradation of chitosan

As described earlier, chitosan is a major component of the walls of the common soil fungi, the zygomycetes, and is produced by deacetylation of chitin to form a major organic component of estuarine sediments. Chitosanase was discovered and shown to be widespread among microbes by Monaghan et al.(1973) and Monaghan (1975). It is produced by bacteria such as species by *Myxobacter, Sporocytophaga, Arthrobacter, Bacillus* and *Streptomyces*, and by fungi such as species of *Rhizopus, Aspergillus, Penicillium, Chaetomium* and the basidiomycete that is a very rich source of glucanase, 'Basidiomycete sp. QM 806'. Davis & Eveleigh (1984) screened soils from barnyard, forest and salt marsh for chitosan-degrading bacteria, and found them at 5.9, 1.5 and 7.4%, respectively of the total heterotrophic isolates, compared with 1.7, 1.2 and 7.4% chitin-degraders. They investigated chitosanase production by a soil isolate of *Bacillus circulans* in more detail, and showed that it was inducible by chitosan but not by chitin or carboxymethylcellulose, and was only active on chitosan. In contrast, the chitosanase from a soil isolate of *Myxobacter* species was active against both chitosan and carboxymethylcellulose (Hedges & Wolfe 1974).

Biotechnology of chitinases and chitosanases

With chitin and chitosan being an enormous renewable resource, much currently going to waste from the shellfish and fungal fermentation industries, and with their essential roles in fungi and invertebrates, it is not surprising that there is much current interest in these polysaccharides and in their degradative enzymes (Muzzarelli & Pariser 1978; Hirano & Tokura 1982; Zikakis 1984; Muzzarelli et al. 1986; Skjak-Braek et al. 1989). Nevertheless the number of successful applications has been disappointedly low. Some of those involving lytic enzymes are dealt with here.

Cloning of chitinase genes

Chitinases from bacteria, fungi and plants have been cloned. Of many bacterial isolates, Monreal & Reese (1969) found *Serratia marcescens* and *Serratia liquefaciens* (*Enterobacter liquefaciens*) to be the most active producers of chitinases. Roberts & Cabib (1982) describe purification of the chitinases, and mutant strains with increased production of chitinase have been produced (Kole & Altosaar 1985; Reid & Ogrydziak 1981). Two chitinases genes *chiA* and *chiB* from random cosmid clones of *S. marcescens* have been inserted into *Escherichia coli*, and then into *Pseudomonas fluorescens* and *Pseudomonas putida*, resulting in four strains of genetically manipulated *Pseudomonas* that have considerable chitinase activities (Suslow & Jones 1988). The rationale to this work is to produce chitinolytic rhizosphere bacteria potentially of value for the biocontrol of soil-borne fungal and nematode diseases of crop plants, as chitin is an essential component of fungal walls and nematode egg cases (Gooday 1990d). In another approach using the same genes, Jones et al. (1986, 1988, Taylor et al. (1987) and Dunsmuir & Suslow (1989) have obtained expression of *chiA* in transgenic tobacco plants, using a range of promoters. These transgenic plants showed increased resistance to the tobacco brown-spot pathogen *Alternaria longipes*. Lund et al. (1989) showed that the *chiA* gene product was secreted by the plant cells in a modified form, and suggest that the bacterial signal sequence is functioning in the plant cells and that the chitinase is *N*-glycosylated through the secretory pathway. Fuchs et al. (1986) have characterized five chitinases in *S. marcescens*, and identified clones from a cosmid library encoding for the *chiA* gene. Their aim was biological control via phylloplane and rhizoplane bacteria. Horwitz et al. (1984) describe attempts at cloning the *Serratia* chitinases into *E. coli*, then back into *S. marcescens* on a high copy number plasmid, to produce a bacterium of value for a bioconversion process to treat shellfish waste. They isolated multiple phage clones encoding both *N*-acetylglucosaminidase and chitinase activity, and suggested that these are linked in a chi operon, which was also suggested by Soto-Gil & Zyskind (1984) in their work towards cloning these genes from *Vibrio harveyi* in *E. coli*. Shapira et al. (1989) have cloned a chitinase gene from *S. marcescens* into *E.coli*, and shown that both the *E.coli* containing the appropriate plasmid and enzyme extracts produced by this strain, have potential for biological control of fungal diseases of plants under greenhouse conditions.

Streptomyces species are well-known producers of active chitinases (Jeuniaux 1963). A chitinase from *S.erythraeus* has been purified and sequenced, with 290 amino acid residues, a molecular weight of 30,4000 and two disulphide bridges, but no homology with other chitinases or lysozymes (Hara et al. 1989; Kamei et al. 1989). A chitinase from *S. plicatus* has been cloned from a DNA library and expressed in *Escherichia coli* (Robbins et al. 1988). The *Streptomyces* chitinase was secreted into the periplasmic space of *E.coli* and its signal sequence was removed. A gene for chitinase from *Vibrio vulnificus* has also been cloned into *E. coli*, and was expressed but not secreted into the medium (Wortman et al. 1986). A gene for chitinase in *Saccharomyces cerevisiae* has been cloned by transforming the yeast with vector plasmids containing a genomic library, and screening for over-producing transformants (Kuranda & Robbins 1988).

Plants produce chitinases as major component of their 'pathogenesis-related proteins' induced following attack by potential pathogens or treatment with ethylene (Mauch & Staehlin 1989). These plant chitinases have antifungal activity (Mauch et al. 1988) that can be greater than that of some bacterial chitinases (Roberts & Selitrennikoff 1988). There is now sufficient information to classify the plant chitinases into at least three structural groups: Class I, basic proteins located primarily in the vacuole, sharing amino-terminal sequence homology with wheat germ agglutinin and hevein;

Class II, acidic, extracellular, having sequence homology with the catalytic domain of Class I, but without the hevein domain; Class III, acid, extracellular, with no homologies to Classes I or II (Payne et al, 1990; Shinshi et al. 1990). Several genes for plant chitinases have been cloned (e.g. Broglie et al. 1986; Payne et al. 1990) and expressed in transgenic plants (Linthorst et al. 1990) with the aim of increasing the plants' resistance to fungal pathogens.

Uses of chitinases and chitosanases

Oligomers of chitin and chitosan have value as fine chemicals and as potential pharmaceuticals (Gooday 1990c). As well as direct hydrolysis of chitin by chitinases, a promising development is the characterization of the transglycosylase activities of these enzymes. Thus Usai et al. (1987, 1990) and Nanjo et al. (1989) describe the use of a chitinase from *Nocardia orientalis* for the interconversion of *N*-acetylglucosamine oligomers, especially to produce hexa-*N*-acetylchitohexose, an oligosaccharide with reported antitumour activity (Suzuki et al. 1986). The transglycosylase activity is favoured by a high substrate concentration and a lowered water activity, i.e. in increasing concentrations of ammonium sulphate. The production of the disaccharide, *N,N'*-diacetylchitobiose, from chitin is described by Takiguchi & Shimahara (1988,1989). They isolated two bacteria, *Vibrio anguillarum* strain E-383a and *Bacillus licheniformis* strain X-Fu, the growth of which in chitin-containing medium resulted in the accumulation of 40 and 46%, respectively conversion of chitin to diacetylchitobiose. Pelletier & Sygusch (1990) and Pelletier et al. (1990) describe the characterization of chitosanases from *Bacillus megaterium*, and their use to assay the degree of deacetylation in samples of chitosan. A direct medical use has been suggested for chitinases in the therapy of fungal diseases, in potentiating the activity of antifungal drugs (Pope & Davies 1979; Orunsi & Trinci 1985). Immunological problems however, probably debar this until anti-iodiotypic antibodies for appropriate chitinases are developed.

Chitinases have found extensive use in the preparation of protoplasts from fungi, a technique of increasing importance in biotechnology (Peberdy 1983). Examples include the chitinases from *Aeromonas hydrophila* subsp. *anaerogenes* (Yabuki et al. 1984) and *Streptomyces olivaceoviridis* (Beyer & Diekmann 1985). Chitosanases are required to make protoplasts from species of the Mucorales (Reyes et al. 1985).

Uses of chitinolytic organisms in biocontrol

As most fungal and invertebrate pests and pathogens have chitin as an essential structural component (Gooday 1990d), chitinase activity could have an important place in the repertoire of mechanisms for biological control. Thus the strongly chitinolytic fungus *Trichoderma harzianum* has good potential for the control of a range of soil-borne plant pathogens (Lynch 1987; Sivan & Chet 1989). Dackman et al. (1989) report that chitinase activity is required for soil fungi to infect eggs of cyst nematodes. Sneh (1981) discusses the use of rhizophere chitinolytic bacteria for biological control. Use of genetic manipulation for the development of organisms with enhanced chitinolytic activities for biological control has been discussed earlier. As well as application of the organisms themselves, there have been reports of biological control by addition of chitin to the soil, presumably as this encourages the growth of chitinolytic microbes which then have a better inoculum potential to infect the soil-borne pathogens and pests, but results currently are variable, and the procedures need further investigation (Gooday 1990a).

Allosamidin and demethylallosamidin

These are antibiotics produced by *Streptomyces* strains, discovered independently by Sakuda et al (1987a) and as metabolite A82516 by Somers et al. (1987), in screens for chitinase inhibitors as potential insecticides. Allosamidin is insecticidal to the silkworm by preventing ecdysis. It does not affect egg hatching of the housefly, but prevents development from larvae to pupae. It has an interesting spectrum of activity, strongly inhibiting chitinases from nematodes and fish, less strongly those of insects and fungi, weakly those of bacteria, and not inhibiting yam plant chitinase (Gooday 1990a, c). Allosamidin is a pseudotrisaccharide, being a disaccharide of *N*-acetylallosamine (until now unknown in nature) linked to a novel aminocyclitol derivative, named allosamizoline (Sakuda et al. 1987b). Demethylallosamidin, a minor cometabolite, has similar activity to allosamidin in inhibiting the silkworm chitinase, but is more inhibitory to the chitinase from *Saccharomyces cerevisiae* (Isogai et al. 1989; Sakuda et al. 1990).

Conclusions

It is clear that the simple definition of chitinase activity, 'hydrolysis of N-acetyl-D-glucosaminide (1–4)-β-linkages in chitin and chitodextrins', belies the complexity and diversity of this group of enzymes. There is increasing awareness of the biological roles and importance of chitin and related glucosaminylglycans, both in nature and technology, and we can look forward to major advances in the next few years.

References

Aruchami M, Sundara-Rajulu G & Gowri N (1986) Distribution of deacetylase in arthropoda. In: Muzzarelli R, Jeuniaux C & Gooday GW (Eds) Chitin in Nature and Technology (pp 263–265). Plenum Press, New York

Baross JA, Tester PA & Morita RY (1978) Incidence, microscopy and etiology of exoskeleton lesions in the tanner crab *Chionectes tanner*. J Fish Res Board Can 35: 1141–1149

Barrett-Bee K & Hamilton M (1984) The detection and analysis of chitinase activity from the yeast form of *Candida albicans*. J Gen Microbiol 130: 1857–1861

Bartnicki-Garcia S (1973) Fundamental aspects of hyphal morphogenesis. In: Ashworth JM & Smith JE (Eds) Microbial Differentiation (pp 245–268). Cambridge University Press, Cambridge

Bartnicki-Garcia S & Lippman E (1982) Fungal wall composition. In Laskin AJ & Lechevalier HA (Eds) CRC Handbook of Microbiology, 2nd Ed. Vol. IV, Microbial Composition: Carbohydrates, Lipids and Minerals (pp 229–252). CRC Press, Boca Raton

Bassler, B., Gibbons P & Roseman S (1989) Chemotaxis to chitin oligosaccharides by *Vibrio furnissii*, a chitinivorous marine bacterium. Biochem Biophys Res Comm 161: 1172–1176

Benhamou N & Asselin A (1989) Attempted localization of a substrate for chitinases in plant cells reveals abundant *N*-acetyl-D-glucosamine residue in secondary walls. Biol Cell 67: 341–250

Berkeley RCW (1979) Chitin, chitosan and their degradative enzymes. In: Berkeley RCW, Gooday GW & Ellwood DC (Eds) Microbial Polysaccharides and Polysaccharases (pp 205–236). Academic Press, London

Beyer M & Diekmann H (1985) The chitinase system in *Streptomyces* sp. ATCC 11238 and its significance for fungal cell wall degradation. Appl Microbiol Biotech 23: 14–146

Bibochka MJ & Khachatourians (1988) *N*-Acetyl-D glucosamine-mediated regulation of extracellular protease in the entomopathogenic fungus *Beauvaris bassiana*. Appl Environ Microbiol 54: 2699–2704

Blackwell J (1988) Physical methods for the determination of chitin structure and conformation. Meth Enzymol 161: 435–442

Borkott H & Insam H (1990) Symbiosis with bacteria enhances the use of chitin by the springtail, *Folsomia candida* (Collembola). Biol Fert Soils 9: 126–129

Bradbury P, Deroux G & Campillo A (1987) The feeding of a chitinivorous ciliate. Tissue Cell 19: 351–363

Broglie KE, Gaynor JJ Broglie RM (1986) Ethylene-regulated gene expression: molecular cloning of the genes encoding an endochitinase from *Phaseolus vulgaris*. Proc Nat Acad Sci USA 86: 6820–6824

Cabib E, Silverman SJ, Sburlati A & Slater ML (1990) Chitin synthesis in yeast *Saccharomyces cerevisiae*. In: Kuhn PJ, Trinci APJ, Jung MJ, Goosey MW & Copping LG (Eds) Biochemistry of Cell Walls and Membranes in Fungi (pp 31–41). Springer-Verlag, Berlin

Chan J G (1970) The Occurrence, Taxonomy and Activity of Chitinolytic Bacteria from Sediment, Water and Fauna of Puget Sound. Ph.D. thesis, University of Washington, Seattle

Dackman C, Chet I & Nordbring-Hertz B (1989) Fungal parasitism of the cyst nematode *Heterodera schachtii*: infection and enzymatic activity. FEMS Microbiol Ecol 62 201–208

Daoust RA & Gunner HB (1979) Microbial synergists pathogenic to *Lymantria dispar*: chitinolytic and fermentative bacterial interactions. J Invert Path 33: 368–377

Datema R, Ende H van den & Wessels JGH (1977) The hyphal wall of *Mucor mucedo* 2. Hexosamine-containing polymers. Eur J Biochem. 80: 621–626

Davis B & Eveleigh DE (1984) Chitosanases: occurrence, production and immobilization. In: Zikakis JP (Ed) Chitin, Chitosan and Related Enzymes (pp 161–179). Academic Press, Orlando

Davis LL & Bartnicki-Garcia S (1984) The co-ordination of chitosan and chitin synthesis in *Mucor rouxii*. J Gen Microbiol 130: 2095–2102

Derksen ACG & Granados RR (1988) Alteration of a lepidoteran peritrophic membrane by baculoviruses and enhancement of viral infectivity. Virology 167: 242–250

Dunsmuir P & Suslow T (1989) Structure and regulation of organ- and tissue-specific-genes: chitinase genes in plants. In: Schell J & Vaszil IK (Eds) Cell Culture and Somatic Cell Genetics in Plants, Vol 6, Molecular Biology of Plant Nuclear Genes (pp 215–227). Academic Press, San Diego

Elango N, Correa JV, Cabib E (1982) Secretory nature of yeast chitinase. J Biol Chem 257: 1398–1400

Flyg C & Boman HG (1988) *Drosphila* genes *cut* and *miniature* are associated with the susceptibility to infection in *Serratia marcescens*. Genet Res 52: 51–56

Fuchs R, McPherson S & Drahos D (1986) Cloning of a *Serratia marcescens* gene encoding chitinase. Appl Environ Microbiol 51: 504–509

Gooday GW (1979) A survey of polysaccharase production: a search for phylogenetic implications. In: Berkeley RCW, Gooday GW & Ellwood DC (Eds) Microbial Polysaccharides and Polysaccharaes (pp 437–460). Academic Press, London

(1990a) The ecology of chitin degradation. In: Marshall KC (Ed) Advances in Microbial Ecology, Vol. 11 (pp 387–430). Plenum Press, New York

(1990b) Inhibition of chitin metabolism. In: Kuhn PJ, Trinci APJ, Jung MJ, Goosey MW & Copping LG (Eds) Biochemistry of Cell Walls and Membranes in Fungi (pp 61–79). Springer-Verlag, Berlin

(1990c) Chitinases. In: Leatham G & Himmel M (Eds) Enzymes in Biomass Conversion. American Chemical Society Books, Washington

(1990d) Chitin metabolism: a target for antifungal and antiparasitic drugs. Pharmacol Ther Suppl 175–185

Gooday GW & Gow NAR (1991) The enzymology of tip growth in fungi. In: Heath IB (Ed) Tip Growth of Plant and Fungal Cells (pp 31–58). Academic Press, New York

Gooday GW, Humphreys AM & McIntosh WH (1986) Roles of chitinase in fungal growth. In: Muzzarelli, RAA, Jeuniaux C & Gooday GW (Eds) (pp 83–91). Plenum Press, New York

Gooday GW, Prosser JI, Hillman K & Cross M (1991) Mineralization of chitin in an estuarine sediment: the importance of the chitosan pathway. Biochem Systematics Ecol (in press)

Gow NAR & Gooday GW (1983) Ultrastructure of chitin in hyphae of *Candida albicans* and other dimorphic and mycelial fungi. Protoplasma 115: 52–58

Hara S, Yamamura Y, Fujii Y, Mega T & Ikenaka T (1989) Purification and characterization of chitinase produced by *Streptomyces erythraeus*. J Biochem 105: 484–489

Hedges A & Wolfe RS (1974) Extracellular enzyme from *Myxobacter AL-1* that exhibits both β-1,4-glucanase and chitosanase activities. J Bacteriol 120: 844–853

Hellebust JA & Lewin J (1977) Heterotrophic nutrition. In: Werner D (Ed) The Biology of Diatoms (pp 169–197). Blackwell, Oxford

Helmke E & Weyland H (1986) Effect of hydrostatic pressure and temperature on the activity and synthesis of chitinases of Antarctic Ocean bacteria. Mar Biol 91: 1–7

Herwig RP, Staley JT, Nerini MK & Braham HW (1984) Baleen whales: preliminary evidence for forestomach microbial fermentation. Appl Environ Microbiol 47 421–423

Hillman K, Gooday GW & Prosser JI (1989a) The mineralization of chitin in the sediments of the Ythan Estuary, Aberdeenshire, Scotland. Estuarine Coastal Shelf Sci 29: 601–612

(1989b) A simple model system for small scale *in vitro* study of estuarine sediment ecosystems. Lett Appl Microbiol 4: 41–44

Hirano S & Tokura S (1982) Chitin and Chitosan. Japanese Society of Chitin and Chitosan. Tottori

Hood MA & Meyers SP (1977) Microbial and chitinoclastic activities associated with *Panaeus setiferus*. J Oceangraph Soc Japan 33 235–241

Horwitz M, Reid J & Ogaydziak D (1984) Genetic improvements of chitinase production of *Serratia marcescens*. In: Zikakis J (Ed) Chitin, Chitosan and Related Enzymes (pp 191–189). Academic Press, Orlando

Humphreys A M & Gooday GW (1984a) Properties of chitinase activities from *Mucor mucedo*: Evidence for a membrane-bound zymogenic form. J Gen Microbiol 130: 1359–1366

(1984b) Phospholipid requirement of microsomal chitinase from *Mucor mucedo*. Curr Microbiol 11: 187–190

(1984c) Chitinase activities from *Mucor mucedo*. In: Nombela C. (Ed) Microbial Cell Wall Synthesis and Autolysis (pp 269–273). Elsevier, Amsterdam

Isaac S & Gokhale AV (1982) Autolysis: a tool for protoplast production from *Aspergillus nidulans*. Trans Br Mycol Soc 78: 389–394

Isogai A, Sato, M, Sakuda S, Nakuyama A (1989) Structure of demethylallosamidin as an insect chitinase inhibitor. Agric Biol Chem 53: 2825–2826

Iten W & Matile P (1970) Role of chitinase and other lysosomal enzymes of *Coprinus lagopus* in the autolysis of fruiting bodies. J Gen Microbiol 61: 301–309

Iverson, KL, Bromel MC, Anderson Aw & Freeman TP (1984) Bacterial symbionts in the sugar beet root maggot *Tetanops myopaeformis* (von Röder), Appl Environ Microbiol 47: 22–27

Janszen FH & Wessels JGH (1970) Enzymic dissolution of hyphal septa in a Basidiomycete. Antonie van Leewenhoek 36: 255–257

Jeuniaux C (1963) Chitine and Chitinolyse. Masson, Paris

(1982) La chitine dans le régne animal, Bull Soc Zool France 107: 363–386

Jones J, Grady K, Suslow T & Bedbrook J (1986) Isolation and characterization of genes encoding two distinct chitinase enzymes from *Serratia marcescens* EMBO J 5: 467–473

Jones JDG, Dean C, Gidoni D, Gilbert D, Bond-Nutter D, Nedbrook J & Dunsmuir P (1988) Expression of a bacterial chitinase protein in tobacco leaves using two photosynthetic gene promoters. Mol Gen Genet 212: 536–542

Kamei, K, Yamamura Y, Hara, S & Ikenda T (1989) Amino acid sequence of chitinase from *Streptomyces erythaeus*. J Biochem 105: 979–985

Kaneko T & Colwell RR (1978) The annual cycle of *Vibrio*

parahaemolyticus in Chesapeake Bay. Microbial Ecol 4: 135–155

Kohlmeyer J (1972) Marine fungi deteroriating chitin of hydrozoa and keratin-like annelid tubes. Mar Biol 12: 277–284

Kole MM & Altosaar I (1985) Increased chitinase production by a non-pigmented mutant of *Serratia marcescens*. FEMS Microbial Lett 26: 265–269

Kuhl J, Nittinger, J & Siebert G (1978) Vervetang von Krillschalen in Futterungsversuchen an de Ratte, Arch Fischereiwiss 29: 99–103

Kuranda MJ & Robbins PW (1988) Cloning and heterologus expression of glycosidase genes from *Saccharomyces cerevisiae*. Proc Nat Acad Sci USA 84: 2585–2589

Lindsay GJH & Gooday GW (1985a) Action of chitinase in spines of the diatom *Thalassiosira fluviatilis*. Carbohydr Polymers 5 131–140

(1985b) Chitinolytic enzymes and the bacterial microflora in the digestive tract of cod, *Gadus morhua*. J Fish Biol 26: 255–265

Lindsay GJH, Walton MJ, Adron, JW, Fletcher, TC, Cho CY & Cowey CB (1984) The growth of rainbow trout (*Salmo gairdneri*) given diets containing chitin and its relationships to chitinolytic enzymes and chitin digestibility. Aquaculture 37: 315–334

Linthorst HJM, Loon LC van, Rossum CMA van, Mayer, A, Bol JF, Roekel JSC van, Meulenhoff EJS & Cornelissen BJC (1990) Analysis of acid and basic chitinases from tobacco and petunia and their constitutive expression in transgenic tobacco. Mol Plant-Microbe Interact (in press)

Lund, P, Lee RY & Dunsmuir P (1989) Bacterial chitinase is modified and secreted in transgenic tobacco. Plant Physiol 91: 130–135

Lynch, J (1987) In vitro identification of *Trichoderma harzianum* as a potential antagonist of plant pathogens. Curr Microbiol 16: 49–53

Lysenko 0 (1976) Chitinase of *Serratia marcescens* and its toxicity to insects, J Invertebr Pathol 27: 385–386

Mauch F, Mauch-Mani B & Boller T (1988) Antifungal hydrolases in pea tissue. Inhibition of fungal growth by combination of chitinase and β-1,3-glucanase. Plant Physiol 87: 936–942

Mauch F & Staehelin LA (1989) Functional implications of the subcellular localization of ethylene-induced chitinase and β-1,3-glucanase in bean leaves. Plant Cell 1: 447–457

Maudlin I & Welburn SC (1988) Parasit Today 4: 109–111

Monaghan RL (1975) The Discovery, Distribution and Utilization of Chitosanase. Ph.D. Thesis, Rutgers University, New Brunswick, N.J.

Monaghan RL, Eveleigh DE, Tewari RP & Reese ET (1973) chitosanase, a novel enzyme, Nature (London) New Biol 245: 78–80

Monreal J & Reese ET (1969) The chitinase of *Serratia marcescens*. Can J Microbiol 15: 689–696

Murray CL & Lovett J S (1966) Nutritional requirements of the chytrid, *Karlingia asterocysta*, an obligate chitinophile. Am J Bot 53: 469–476

Muzzarelli RAA & Pariser ER (1978) Proceedings of the First International Conference on Chitin/Chitosan. MIT Sea Grant Report MITSG

Muzzarelli RAA, Jeuniaux C & Gooday GW (1986) Chitin in Nature and Technology. Plenum, New York

Nanjo F, Sakai K, Ishikawa M, Isobe K & Usui T (1989) Agric Biol Chem 53: 2189–2195

Orunsi NA & Trinci APJ (1985) Growth of bacteria on chitin, fungal cell walls and fungal biomass, and the effect of extracellular enzymes produced by these cultures on the antifungal activity of amphotericin B. Microbios 43: 17–30

Patton RS & Chandler, PT (1975) In vivo digestibility evaluation of chitinous material. J Dairy Sci 58: 397–403

Payne S, Ahl P, Moyer, M, Harper, A, Beck J, Meins F & Ryals J (1990) Isolation of complementary DNA clones encoding pathogeneis-related proteins P and Q, two acid chitinases from tobacco. Proc Nat Acad Sci USA 87: 98–102

Peberdy J (1983) Genetic recombination in fungi following protoplast fusion and transformation. In: Smith JE (Ed) Fungal Differentiation (pp 559–581). Dekker, New York

Pel R & Gottschal JC (1986a) Chitinolytic communities from an anaerobic estuarine environment, In Muzzarelli RAA, Jeuniaux C & Gooday GW (Eds) Chitin in Nature and Technology (pp 539–546). Plenum, New York

(1986b) Stimulation of anaerobic chitin degradation in mixed cultures, Antonie van Leewenhoek 52: 359–360

(1989) Interspecies interaction based on transfer of a thioredoxin-like compound in anaerobic chitin-degrading mixed cultures. FEMS Microbiol Ecol 62: 349–358

Pel R, Wessels G, Aalfs H & Gottschal JC (1989) Chitin degradation by *Clostridium* sp. strain 9.1 in mixed cultures with saccharolytic and sulphate-reducing bacteria. FEMS Microbiol Ecol 62: 191–200

Pel R, Wijngaard Aj Van Den, Epping E & Gottschal JC (1990) Comparison of the chitinolytic properties of *Clostridium* sp. strain 9.1 and a chitin degrading bacterium from the intestinal tract of the plaice *Pleuronectes platessa* (L.). J Gen Microbiol 136: 695–704

Pelletier A & Sygusch J (1990) Purification and characterization of three chitosanase activities from *Bacillus megaterium* P1. Appl Environ Microbiol 56: 844–848

Pelletier A, Lemire I, Sygusch J, Chornet E & Overend RP (1990) Chitin/chitosan transformation by thermo-mechanochemical treatment including characterization by enzymatic depolymerisation. Biotech Bioeng 36: 310–315

Peter MG, Kegel G & Keller R (1986) Structural studies on sclerotized insect cuticle. In: Muzzarelli R, Jeuniaux C, Gooday GW (Eds) Chitin in Nature and Technology (pp 21–28). Plenum Press, New York

Petz W, Foissner W, Wirnsberger E, Kruatgartner Wd & Adam H (1986) Mycophagy, a new feeding strategy in autochthonous ciliates. Naturwissenschaften 73: S.560–561

Pope, AMS & Davies DAL (1979) The influence of carbohydrases on the growth of fungal pathogens in vitro and in vivo. Postgrad Med J 55: 674–676

Poulicek, M, Voss-Foucart MF & Jeuniaux C (1986) Chitinoproteic complexes and mineralization in mollusk skeletal structures. In: Muzzarelli R, Jeuniaux C, Gooday GW (Eds) Chitin in Nature and Technology (pp 7–12). Plenum Press, New York

Reid, JD & Ogrydziak DM (1981) Chitinase-overproducing mutant of *Serratia marcescens*. Appl Environ Microbiol 41: 664–669

Reichardt W, Gunn, B and Colwell RR (1983) Ecology and taxonomy of chitinoclastic *Cytophaga* and related chitin-degrading bacteria isolated from an estuary, Microb Ecol 9: 273–294

Reisert PS & Fuller MS (1962) Decomposition of chitin by *Chytridiomyces* species. Mycologia 54: 647–657

Reyes, F, Perez-Leblic MI, Martinez MJ & Lahoz R (1984) Protoplast production from filamentous fungi with their own autolytic enzymes. FEMS Microbiol Lett 24: 281–283

Reyes, F, Lahoz R, Martinez MJ & Alfonso C (1985) Chitosanases in the autolysis of *Mucor rouxii*. Mycopathologia 89: 181–187

Robbins PW, Albright C & Benfield B (1988) Cloning and expression of a *Streptomyces plicatus* chitinase (chitinase-63) in *Escherichia coli*. J Biol Chem 263: 443–447

Roberts RL & Cabib E (1982) *Serratia marcescens* chitinase: one-step purification and use for the determination of chitin. Anal Biochem 127: 402–412

Roberts WK & Selitrennikoff CP (1988) Plant and bacterial chitinases differ in antifungal activity. J Gen Microbiol 134: 169–176

Sakuda S, Isogai A, Matsumoto S & Suzuki A (1987a) Search for microbial insect growth regulators II. Allosamidin, a novel insect chitinase inhibitor. J Antibiot 40: 296–300

Sakuda S, Isogai A, Makita T, Matsumoto S, Koseki K, Kodama H & Suzuki A (1987b) Structures of Allosamidins, novel insect chitinase inhibitors, produced by actinomycetes. Agric Biol Chem 51: 3251–3259

Sakuda A, Nishimoto Y, Ohi M, Watanabe M, Takayama J, Isogai A & Yamada Y (1990) Effects of demethylallosamidin, a potent yeast chitinase inhibitor, on the cell division of yeast. Agric Biol Chem 54: 1333–1335

Sassen MMA (1965) Breakdown of the plant cell wall during the cell fusion process. Acta Bot Neerlandica 14: 165–196

Seki H & Taga N (1965) Microbial studies on the decomposition of chitin in marine environment-VI. Chitinoclastic bacteria in the digestive tract of whales from the Antarctic Ocean. J Oceanogr. Soc Japan 20: 272–277

Shapira R, Ordentlich A, Chet I & Oppenheim AB (1989) Control of plant diseases by chitinase expressed from cloned DNA in *Escherichia coli*. Phytopathology 79: 1246–1249

Shinshi H, Beuhaus J, Ryals J & Meins F (1990) Structure of a tobacco chitinase gene: evidence that different chitinase genes can arise by a transposition of sequences encoding a cysteine-rich domain. Plant Mol Biol 14: 357–368

Sietsma, JH, Vermeulen CA & Wessels JGH (1986) The role of chitin in hyphal miorphogenesis. In Muzzarelli R, Jeuniaux C & Gooday GW (Eds) Chitin in Nature and Technology (pp 6369). Plenum Press, New York

Sivan A & Chet I (1989) Degradation of fungal cell walls by lytic enzymes of *Trichoderma harzianum*. J Gen Microbiol 135: 675–682

Skjak-Braek G, Anthonsen T & Sandford P (1989) Chitin and Chitosan. Elsevier, London

Smith RJ & Grula EA (1983) Chitinase is an inducible enzyme in *Beauvaria bassiana*. I Invertebr Path 42 319–326

Sneh B (1981) The use of rhizophere chitinolytic bacteria for biological control. Phytopath Zeitscrift 100: 251–256

Soderhall K & Unestam T (1975) Properties of extracellular enzymes from *Aphanomyces astaci* and their relevance in the penetration process of crayfish cuticle. Physiol Plant 35: 140–146

Somers PJB, Yao RC, Doolin LR, McGowan MJ, Fakuda DS & Mynderse JS (1987) Method for the detection and quantitation of chitinase inhibitors in fermentation broths: Isolation and insect life cycle. Effect of A82516. J Antibiot 40: 1751–1756

Soto-Gil RW & Zyskind JW (1984) Cloning of *Vibrio harveyi* chitinase and chitobiase gene in *Escherichia coli*. In: Zikakis JP (Ed) Chitin, Chitosan and Related Enzymes (pp 209–223). Academic Press, Orlando

Srivastava AK, Defago G & Boller T (1985) Secretion of chitinase by *Aphanocladium album*, a hyperparsite of wheat. Experientia 41: 1612–1613

St Leger RJ, Cooper RM & Charnley AK (1986) Cuticle-degrading enzymes of entomopathogenic fungi: regulation of production of chitinolytic enzymes. J Gen Microbiol 132: 1509–1517

Surarit, R, Gopel PK & Shepherd MG (1988) Evidence for a glycosidic linkage between chitin and glucan in the cell wall of *Candida albicans*. J Gen Microbiol 134: 1723–1730

Suslow TV & Jones J (1988) Chitinase-producing bacteria. US Patent No 4751081

Suzuki K, Mikami T, Okawa Y, Tokora A, Suzuki S & Suzuki M (1986) Antitumour effect of hexa-*N*-acetylchitohexase and chitohexaose. Carb Res 151: 403–408

Takiguchi Y & Shimahara K (1988) N,N-Diacetylchitobiose production from chitin by *Vibrio anguillarum* strain E-383a. Lett Appl Microbiol 6 129–131

(1989) Isolation and identification of a thermophilic bacterium producing *N,N*-diacetylchitobiose from chitin. Agric Biol Chem 53: 1537–1541

Taylor JL, Jones JDG, Sandler S, Mueller GM, Bedbrook J & Dunsmuir P (1987) Optimizing the expression of chimeric genes in plant cells. Mol Gen Genet 210: 572–577

Tracey MV (1955) Chitinase in some Basidiomycetes. Biochem J 61: 579–589

Usai T, Hayashi Y, Nanjo F, Sakai K & Ishido Y (1987) Transglycosylation reaction of a chitinase purified from *Nocardia orientalis*. Biochem Biophys Acta 923: 302–309

Usai T, Matsui M & Isobe K (1990) Enzymic synthesis of useful chito-oligosaccharides utilizing transglycosylation by chitinolytic enzymes in buffer containing ammonium sulfate. Carb Res 202: 65–78

Walker, AN, Garner RE & Horst MN (1990) Immunocytochemical detection of chitin in *Pneumocystis carinii*. Infect Immun 58: 412–415

Wessels JGH (1988) A steady state model for apical wall growth. Acta Bot Neerl 37:3–16

Wolkin, RH & Pate JKL (1985) Selection for nonadherent or nonhydrophobic mutants co-selects for non-spreading mutants of *Cytophage johnsonae* and other gliding bacteria. J Gen Microbiol 131: 7737–750

Wortman AT, Somerville CC & Colwell RR (1986) Chitinase determinants of *Vibrio vulnificus*: Gene cloning and applications of a chitinase probe. Appl Environ Microbiol 52: 142–145

Yabuki M, Kasai Y, Ando A & Fujii T (1984) Rapid method for converting fungal cells into protoplasts with a high regeneration frequency. Exp Mycol 8: 386–390

Zarain-Herzberg A & Arroya-Begovich A (1983) Chitinolytic activity from *Neurospora crassa*. J Gen Microbiol 129: 3319–3326

Zhloba NM, Tiunova NA & Sidorova II (1980) extracellular hydrolytic enzymes of mycophilic fungi. Mikol Fitopatol 14: 496–499

Zikakis J (1984) Chitin, chitosan and Related Enzymes. Academic Press, Orlando

Biodegradation **1**: 191–206, 1990.

The biodegradation of aromatic hydrocarbons by bacteria

Mark R. Smith
Division of Industrial Microbiology, Agricultural University, P.O. Box 8129, 6700 EV Wageningen, The Netherlands

Key words: alkenylbenzenes, alkylbenzenes, arenes, benzene, biphenyl, fused aromatic compounds, single bacterial isolates

Abstract

Aromatic compounds of both natural and man-made sources abound in the environment. The degradation of such chemicals is mainly accomplished by microorganisms. This review provides key background information but centres on recent developments in the bacterial degradation of selected man-made aromatic compounds. An aromatic compound can only be considered to be biodegraded if the ring undergoes cleavage, and this is taken as the major criteria for inclusion in this review (although the exact nature of the enzymic ring-cleavage has not been confirmed in all cases discussed).

The biodegradation of benzene, certain arenes, biphenyl and selected fused aromatic hydrocarbons, by single bacterial isolates, are dealt with in detail.

Introduction

For the purposes of this review aromatic compounds are restricted to benzene and compounds that resemble benzene in chemical structure. Benzene and related compounds are characterised by their possession of a large (negative) resonance energy. This results in a thermodynamic stability which manifests itself in chemical properties very different from those observed for aliphatic (including alicyclic) compounds. However, most of the hydrocarbons discussed here fall more correctly into the category of compounds classified as arenes. In this review the term 'aromatic' will be used for those hydrocarbon compounds containing an aromatic moiety which ultimately undergoes enzymatic ring-cleavage.

Aromatic hydrocarbons are ubiquitous in nature. Although there is some debate as to their origin in the environment, it is generally accepted that most are not of biosynthetic origin but are derived from the (natural) pyrolysis of organic compounds (Gibson & Subramanian 1984). Indeed, next to glucosyl residues, the benzene ring is the most widely distributed unit of chemical structure in nature (Dagley 1981). It is of little surprise therefore that micro-organisms have evolved capable of degrading aromatic compounds. Today, there is great concern regarding the occurrence of man-made aromatic hydrocarbons in the environment. Benzene, toluene, ethylbenzene, styrene and the xylenes are among the 50 largest-volume industrial chemicals produced, with production figures of the order of millions of tons per year. These compounds are widely used as fuels and industrial solvents. In addition, they and polynuclear aromatic compounds provide the starting materials for the production of pharmaceuticals, agrochemicals, polymers, explosives and many other everyday products (Gibson 1971).

The use of man-made aromatic hydrocarbons has inevitably led to their release (either accidental or otherwise) into the environment and this problem is still escalating in spite of governmental intervention.

The biodegradation of aromatic compounds can be considered, on the one hand as part of the normal process of the carbon cycle, and as the removal of man-made pollutants from the environment, on the other.

Over the last decades this topic has been extensively reviewed (Gibson 1971; Hopper 1978; Cripps & Watkinson 1978; Gibson & Subramanian 1984; Cerniglia 1984; Dagley 1986, as examples). Since the last extensive review there have been many advances in the field at the physiological, biochemical and molecular biological levels. The majority of the recent advances made have consolidated previous findings giving a fuller, if not complete, picture.

It is impossible in such a review article to cover all the aspects of bacterial aromatic degradation of the last few years and this therefore is more of a personal view on recent developments in certain areas. It is not the intention to provide an historical account of the developments in this area as this has been more than adequately been achieved by others (Gibson & Subramanian 1984). Instead it hoped to provide the reader with some key developments in the defined areas (although some background information will be provided) and to raise some questions regarding the current inadequacies in our knowledge.

The parent member of the aromatic hydrocarbons is benzene and it is therefore logical to begin by considering the biodegradation of the 'begetter' of the other members of the series.

Benzene

There have been few reports on the bacterial biodegradation of benzene in the literature over the last five years. The excellent studies carried out in the previous three decades (Marr & Stone 1961; Gibson et al. 1968 Gibson et al. 1970; Högn & Jaenicke 1972; Axell & Geary 1975) elucidated the pathways involved, identified the intermediates and characterised the enzyme systems. Figure 1 shows the two divergent pathways employed. Both share the same initial mode of attack resulting in the formation of catechol which is further catabolised by either catechol 1,2-dioxygenase (the so called *ortho*- or intradiol-cleavage) and subsequently via the β-ketoadipate pathway or catechol 2,3-dioxygenase (the so called *meta*- or extradiol-cleavage). Both routes have been described in different benzene utilising strains.

Although there have been recent reports of the isolation of new bacterial strains which can grow on benzene (Shirai 1986; van den Tweel et al. 1988; Winstanley et al. 1987, as examples) the biodegradation routes were, not surprisingly, the same as those outlined above (Fig. 1).

Winstanley et al. (1987) described a new benzene utilising bacterium, *Acinectobacter calcoaceticus* RJE74, which carries a large plasmid (pWW174) encoding the enzymes for the catabolism of benzene via the β-ketoadipate pathway (that involving catechol 1,2-dioxygenase). This was the first report of the *ortho* pathway being plasmid encoded and was one of very few citations of plasmids in this genus.

The biodegradation of benzene has also been re-examined from a biotechnological standpoint. Interest has focused on the production of the first two intermediates in the pathways (*cis*-benzeneglycol, CBG and catechol). CBG formed biologically from benzene has been reported by various groups (Ballard et al. 1983; Ley et al. 1987; van den Tweel et al. 1988). The product (generated by mutant strains of benzene utilisers) is projected to find uses as a chiral building block for polymers and pharmaceuticals. Shirai (1986) selected a mutant strain of *Pseudomonas* sp. capable of accumulating catechol and subsequently (Shirai 1987) investigated the possibility of the use of free and immobilised cells for an industrial process. Catechol and its derivatives are important chemicals used mainly for the production of synthetic flavours such as vanillin.

Arenes

The introduction of a substituent group(s) onto the benzene ring opens up the possibility of alternative

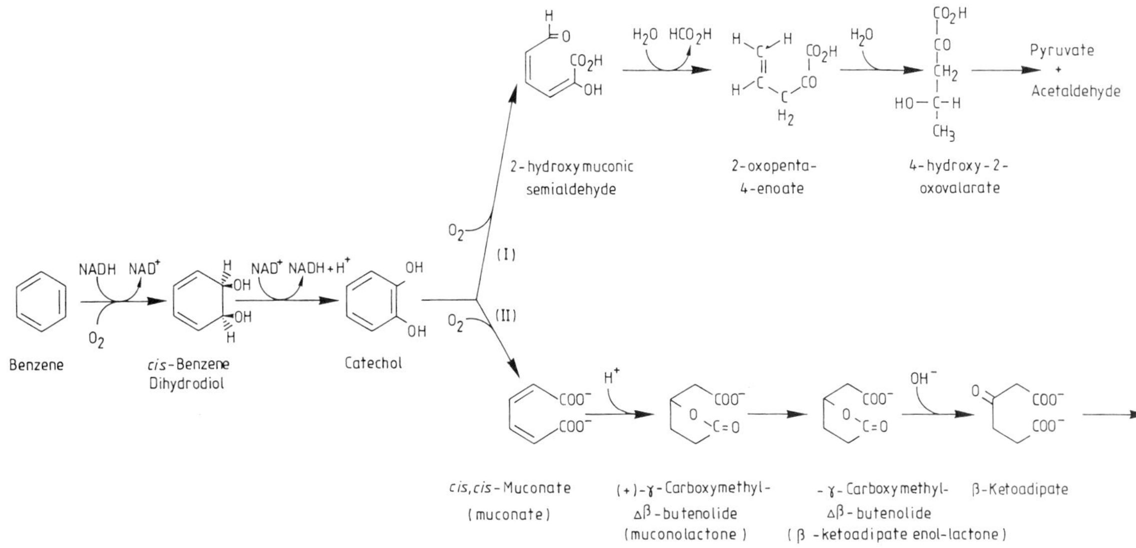

Fig. 1. The biodegradative routes of benzene. (I) *meta*-cleavage route. (II) *ortho*-cleavage route.

modes of biodegradation; either side-chain attack or ring attack. Indeed with the longer chain length alkylbenzenes the oxidation of the side chain is sufficient to support growth and the organisms may not be able to degrade the aromatic moiety. Such compounds may be regarded as substituted alkanes rather than substituted benzenes. For the purposes of this review, I will only discuss those cases where the benzene ring undergoes ring-cleavage (either before or after side chain modifications).

The biodegradation of mono-alkylbenzenes

Toluene, the simplest of these substituted benzenes, is biodegraded by both ring attack and methyl-group hydroxylation. Figure 2 gives the alternative pathways. The evidence for these routes is well established (see Hopper 1978 for review).

Over the last five years, there has been an enormous number of papers on the plasmid-encoded biodegradation of toluene. Williams and his colleagues at the University of Wales (Bangor, UK) have shown that toluene (and *m*-, *p*- xylenes, etc) is degraded via catechol and subsequently the *meta* pathway by several strains of *Pseudomonas* by enzymes encoded on plasmids, designated TOL (Bayly & Barbour 1984). These plasmids often contain two catabolic operons (Nakazawa et al. 1980; Franklin et al. 1981). The 'upper' pathway operon encodes enzymes for the successive oxidation of the hydrocarbons to the corresponding alcohol, aldehyde and carboxylic acid derivatives. The 'lower' or *meta*-cleavage pathway operon encodes enzymes for the conversion of the carboxylic acids to catechols, whose aromatic rings are then cleaved (*meta*-fission) to produce corresponding semialdehydes, which are then further catabolised through the TCA cycle (Ramos et al. 1987). Figure 3 outlines the TOL plasmid encoded pathway.

To cover the huge amount of data published in this area would require more space than is available in such a review. In a recent mini-review Burlage et al. (1989) discussed in detail the most studied of the TOL plasmids (pWWO), however it is the opinion of this author that an up to date detailed account of all of the developments in this field is now needed.

Other recent advances in the study of the biodegradation of toluene include: isolation of a novel strain (Simpson et al. 1987), studies of the growth parameters of *Pseudomonas putida* in chemostat cultures (Vecht et al. 1988) and its anaerobic biodegradation (Zeyer et al. 1990; Lovley & Lonergan 1990).

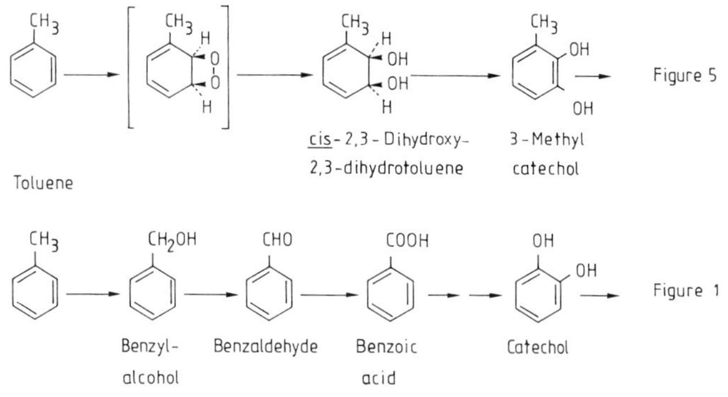

Toluene

cis-2,3-Dihydroxy-2,3-dihydrotoluene

3-Methyl catechol

Figure 5

Benzyl-alcohol

Benzaldehyde

Benzoic acid

Catechol

Figure 1

Fig. 2. The biodegradation routes of toluene.

A thermotolerant *Bacillus* sp. which grew on toluene at 50° C was isolated by Simpson et al. (1987). Biodegradation (Simpson et al. 1987) was via *cis*-toluene dihydrodiol, 3-methylcatechol and *meta*-cleavage (Fig. 2). The *cis*-toluene dihydrodiol dehydrogenase from this organism was purified and shown to possess different properties to those previously reported, most notable was a temperature optimum of 80° C (Simpson et al. 1987).

Vecht et al. (1988) recently reported on the growth of *Pseudomonas putida* in chemostat cultures with toluene as the sole source of carbon and energy (after initial growth on *m*-toluic acid). Steady states were maintained for several months with maximal biomass concentrations of 3.2 g cell dry wt/l. The maximum specific growth rate was $0.13\,h^{-1}$, with a cellular yield of 1.05 g cell dry weight/g toluene utilised.

The anaerobic biodegradation of toluene has been reported by two independent groups. A *Pseudomonas* sp. has been isolated which oxidises toluene to carbon dioxide with NO_3 or N_2O as the potential electron acceptors (Zeyer et al. 1990). It was not demonstrated whether this bacteria could obtain energy from toluene oxidation and the reduction products of nitrate and nitrous oxide were not investigated. Lovley & Lonergan (1990) isolated an unidentified bacterial strain (GS-15) which coupled the oxidation of aromatic compounds (including toluene) to the reduction of Fe(III). This was the first conclusive report of the anaerobic oxidation of an aromatic hydrocarbon. Biodegradation proceeds via the oxidation of the methyl

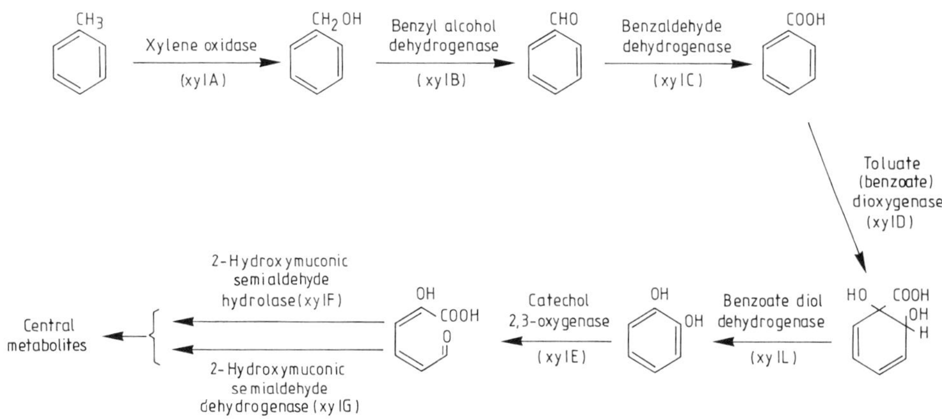

Fig. 3. The early enzymes of the TOL plasmid degradative pathway.

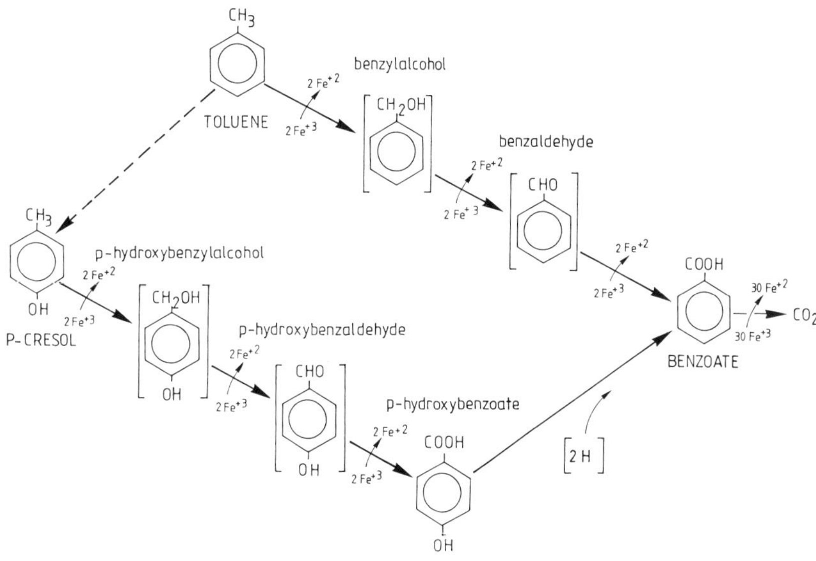

Fig. 4. The anaerobic biodegradation of toluene by strain GS-15.

group (leading to benzoate) or possibly via *p*-cresol, *p*-hydroxybenzoate and benzoate. The proposed pathway is given in Fig. 4. The anaerobic biodegradation of aromatic acids and phenols is well known (Berry et al. 1987; Evans & Fuchs 1988) and it will be interesting to see if aromatic hydrocarbons other than toluene can be anaerobically dissimilated by pure bacterial strains.

Direct cleavage of the aromatic moiety of alkylbenzenes, without prior attack of the alkyl side chain, has been demonstrated for ethylbenzene (Gibson et al. 1973), 2-phenylbutane, 3-phenylpentane (Baggi et al. 1972), n-butylbenzene (Jigami et al. 1974a), isopropylbenzene (Jigami et al. 1975; Eaton & Timmis 1986) and *tert*-butylbenzene (Catelani et al. 1977).

We recently re-examined the biodegradation of alkylbenzenes (Smith & Ratledge 1989a). *Pseudomonas* sp. NCIB 10643 grew on a range of n-alkylbenzenes (C2–C7) and on several branched species within this chain size (isopropylbenzene, isobutylbenzene, *sec*-butylbenzene, *tert*-butylbenzene and *tert*-amylbenzene). All of the alkylbenzenes were catabolised via ring attack, rather than side chain attack, proceeding via initial dioxygenase activity resulting in the corresponding 2,3-dihydro-2,3-dihydroxyalkylbenzene which underwent reduction to the corresponding 2,3-dihydroxyl- intermediate (3-alkylsubstituted catechols). The 3-substituted catechols were ring-cleaved by an extra-diol type enzyme between Cl and C2 resulting in characteristic *meta* ring-fission products. Further catabolism was by hydrolytic attack to give alkyl-chain dependent carboxylic acids and, presumably, 2-oxopenta-4-enoate.

A general pathway for the complete catabolism of these mono-alkylbenzenes is given in Fig. 5.

There have been several reports that the en-

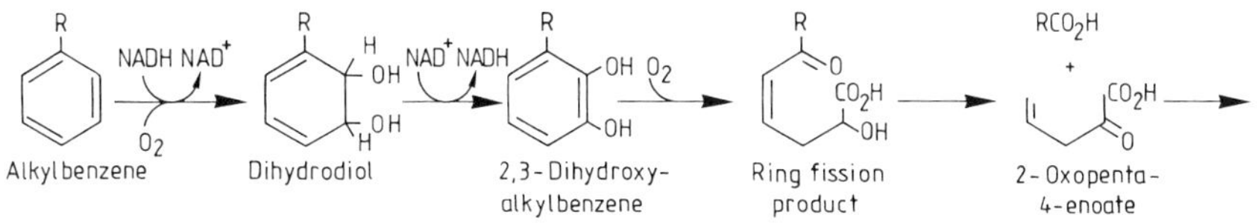

Alkylbenzene — NADH NAD⁺, O2 → Dihydrodiol — NAD⁺ NADH → 2,3-Dihydroxy-alkylbenzene — O2 → Ring fission product → RCO2H + 2-Oxopenta-4-enoate →

Fig. 5. The biodegradation of alkylbenzenes (Cl–C7).

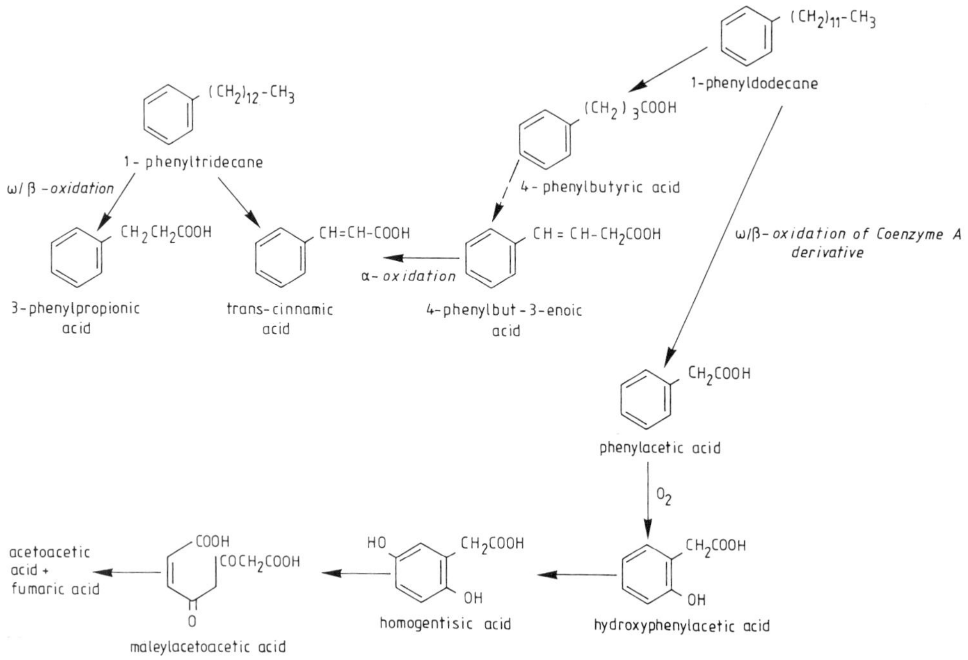

Fig. 6. The biodegradative routes of 1-phenyltridecane and 1-phenyldodecane.

zymes encoding for the biodegradation of alkylbenzenes are plasmid-borne. Bestetti & Galli (1984) showed that the genes for the catabolism of ethylbenzene (and 1-phenylethanol) in *Pseudomonas fluorescens* were sited on a 253–267 kilobase plasmid. Eaton & Timmis (1986) have demonstrated that the catabolism of isopropylbenzene by *Pseudomonas putida* RE 204 is plasmid encoded.The pathway was shown to be identical to that outlined above (see Fig. 5). Our own studies with *Pseudomonas* sp. NCIB 10643 suggested that the genes of alkylbenzene (and biphenyl) biodegradation were chromosomal (Smith & Ratledge 1989b).

When the alkyl chain length exceeds C7 the preferred route seems to be by attack on the alkyl chain. Sariaslani et al. (1974) reported on the degradation of n-dodecyl-and n-nonyl- benzenes (1-phenyldodecane and 1-phenylnonane respectively) by initial side chain attack via ω- and β-oxidation. Both compounds could be catabolised through 2,5-dihydroxyphenylacetic acid (homogentisic acid) and then ring cleaved. Amund & Higgins (1985) later confirmed this route for the degradation of 1-phenyldodecane by *Acinetobacter lwoffi* and also showed that 1-phenyltridecane was transformed to *trans*-cinnamic acids and 3-phenylpropionic acid which were not further metabolised. The proposed scheme for the biodegradation of 1-phenyldodecane and 1-phenyltridecane is given in Fig. 6.

The biodegradation of di-alkylbenzenes

Until recently only the *m*- and *p*- isomers of xylene had been shown to be biodegraded by bacteria. Both compounds are degraded by certain strains of *Pseudomonas* (notably those containing the TOL plasmid – see above) by initial oxidation of one of the methyl groups to the corresponding methyl benzylalcohols, tolualdehydes, toluic acids and methyl catechol (Davey & Gibson 1974; Davis et al. 1968). The biodegradation of *m*- and *p*-xylenes to their corresponding methylcatechols is shown in Fig. 7. The resultant catechols then undergo *meta*-cleavage. The ring-fission products of the two different methyl catechols (3-methylcatechol from *m*-xylene; 4-methylcatechol from *p*-xylene) are catabolised by different enzyme systems (Duggleby &

Williams 1986). The product from 3-methylcatechol cleavage is further degraded by a single hydrolase type enzyme (Duggleby & Williams 1986; Smith & Ratledge 1989a), whereas the product from 4-methylcatechol, an aldehyde, is converted via the enzymes of the 4-oxocrotonate branch (Sala-Trepat et al. 1972; Wigmore et al. 1974). These pathways are illustrated in Fig. 8.

An alternative mode of attack of *p*- and *m*-xylenes is via direct dioxygenase attack of the aromatic moiety yielding the corresponding *cis*-dihydrodiol with subsequent conversion to substituted catechols (3,6-dimethylcatechol from *p*-xylene; 3,5-dimethylcatechol from *m*-xylene) by dehydrogenase type enzymes (Gibson et al. 1974). However, although this is often cited as an alternative pathway for the degradation of xylenes (Baggi et al. 1987, for example) the resultant catechols are not further degraded and this route should be regarded as a biotransformation reaction.

Members of the genus *Nocardia* can co-metabolise all three of the isomers of xylene (see Gibson & Subramanian 1984 for review). Noteworthy, the *p*- and *m*-xylenes were co-metabolised (hexadecane as substrate) via *ortho* cleavage whereas *o*-xylene was attacked by *meta*-cleavage.

The first reports of the complete biodegradation of *o*-xylene as sole source of carbon and energy by pure cultures were provided by Baggi et al. (1987) and Schraa et al. (1987). Initial studies (using a strain of *Pseudomonas stutzeri*) suggested that *o*-xylene was catabolised via 3,4-dimethylcatechol with subsequent *meta*-cleavage (Baggi et al. 1987). Independently, Schraa et al. (1987) reported on the characterisation of *Corynebacterium* strain C125 able to grow on *o*-xylene as the sole source of carbon and energy. The proposed pathway was the same as that suggested by Baggi et al. (1987) and confirmed 2-dihydroxy-5-methyl-6-oxo-2,4-heptadienoate as the ring-fission product. The pathway is illustrated in Fig. 9.

The observation that none of the *p*- and *m*-xylene degrading bacteria can attack *o*-xylene (and vice versa from Baggi et al. 1987; Schraa et al. 1987) raises interesting questions regarding the evolution of these bacterial strains.

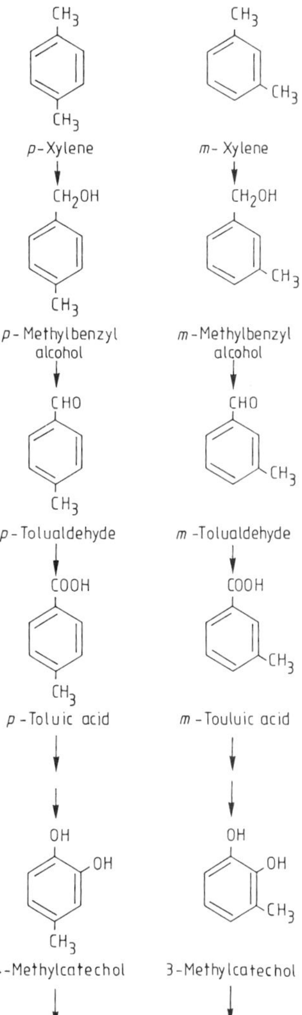

Fig. 7. The initial reactions in the bidegradation of *m*- and *p*-xylenes.

With regard to other di-alkylsubstituted benzenes, there have been no significant advances since the last review (Gibson & Subramanian 1984) was published and our knowledge is still limited to the biodegradation routes of *p*-cymene (DeFrank & Ribbons 1976; DeFrank & Ribbons 1977a, b) and 3-ethyltoluene (Jigami et al. 1974b; Kunz & Chapman 1981). In both these cases, attack begins on the smaller substituent of the ring (i.e. the methyl group), however, there is too little evidence available to judge whether this is always the case.

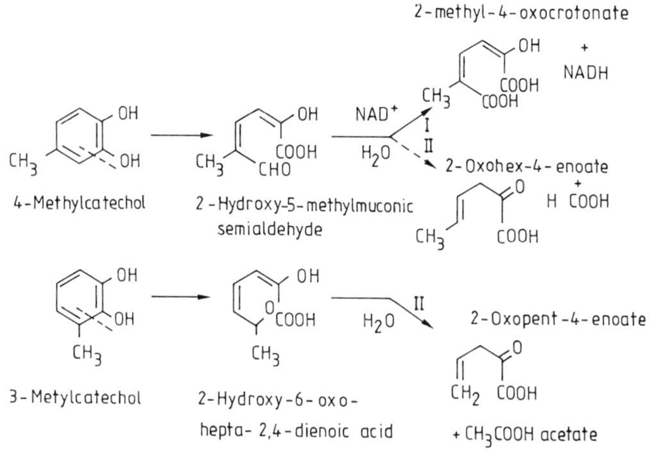

2-methyl-4-oxocrotonate

NADH

4-Methylcatechol 2-Hydroxy-5-methylmuconic semialdehyde 2-Oxohex-4-enoate H COOH

3-Metylcatechol 2-Hydroxy-6-oxo-hepta-2,4-dienoic acid 2-Oxopent-4-enoate + CH_3COOH acetate

Fig. 8. The alternative pathways for the catabolism of 4- and 3-methylcatechols. (I) 4-Oxocrotonate branch. (II) Hydrolytic branch.

The biodegradation of alkenylbenzenes

Reports on the biodegradation of alkenylbenzenes are extremely scarce. This is somewhat surprising when one considers the enormous quantities styrene (the simplest member of this series) produced by the petrochemical industry (3.6 million tons in the United States in 1987 – Hartmans et al. 1989). To date there have been no detailed examinations of the complete degradation of any of the alkenylbenzenes and only a handful on the initial attack. There are however reports of bacteria able to grow on styrene (Baggi et al. 1983; Shirai & Hisatsuka 1979; van den Tweel et al. 1986) and methylstyrenes (Omori et al. 1974; Dzhusupova et al. 1985) as sole sources of carbon and energy and thus by implication these organisms must be capable of ring-cleavage.

The initial degradation of styrene by *Xanthobacter* strain 124X (a strain isolated by enrichment on styrene by van den Tweel et al. 1986) was recently reported by Hartmans et al. (1989). Although styrene oxide and 2-phenylethanol have been implicated in its degradation by other bacteria (Shirai & Hisatsuka 1979) and were shown to be oxidised by cells grown on styrene, these authors concluded that the initial step in styrene metabolism is oxygen dependent and probably involves oxidation of the aromatic nucleus. Previously, this organism was also shown to grow on ethylbenzene and toluene (van den Tweel et al. 1986).

Subsequently, this same group (Hartmans et al. 1990) isolated 14 strains of bacteria able to grow on styrene as the sole source of carbon and energy. One of these (S5) was studied in more detail and the initial reactions characterised. Styrene was converted to styrene oxide by a novel flavin monooxygenase. Further degradation proceeded via phenylacetaldehyde and phenylacetic acid. The conversion of phenylacetic acid to central metabolites was not investigated.

The biodegradation of other members of this class of aromatic hydrocarbons has received very little attention in the last few years. The biotransformation of styrenes by a strain of *Pseudomonas putida* capable of growth on α-methylstyrene has

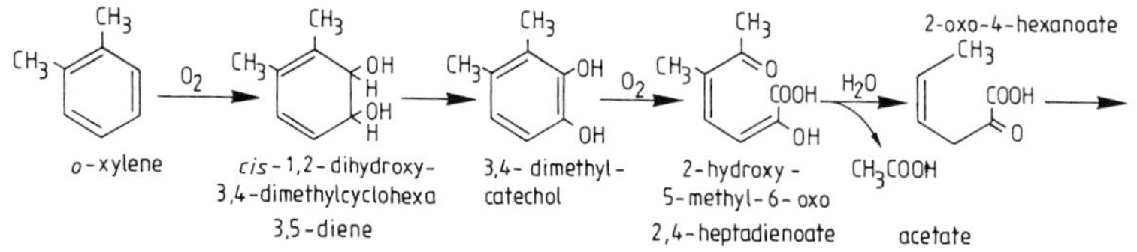

o-xylene — *cis*-1,2-dihydroxy-3,4-dimethylcyclohexa 3,5-diene — 3,4-dimethyl-catechol — 2-hydroxy-5-methyl-6-oxo 2,4-heptadienoate — 2-oxo-4-hexanoate; CH_3COOH acetate

Fig. 9. The biodegradative pathway of *o*-xylene.

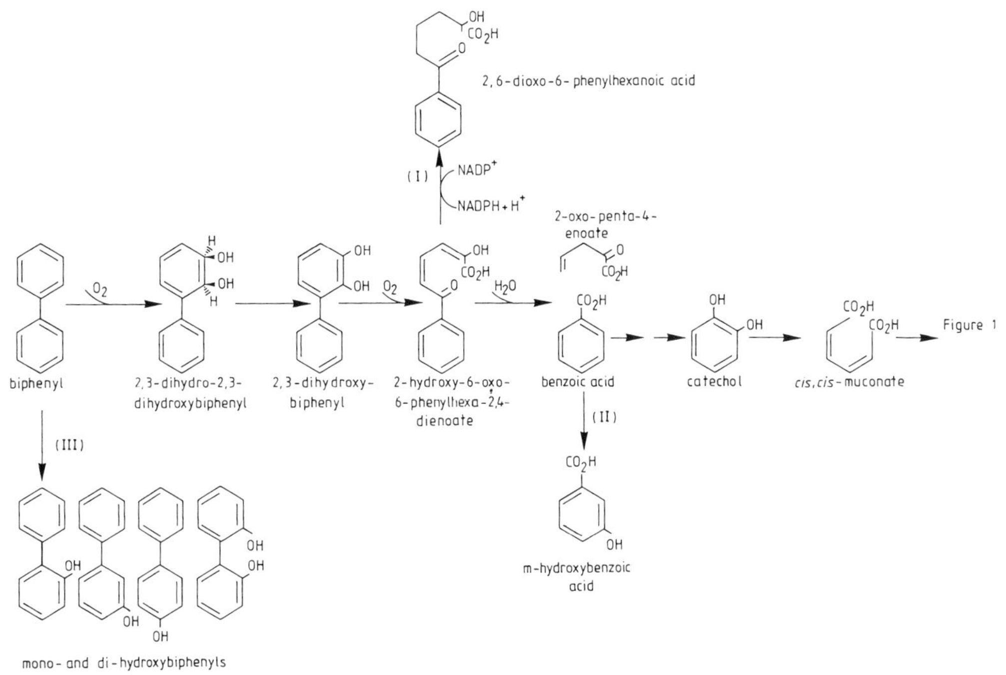

Fig. 10. The major pathway for the biodegradation of biphenyl and some alternative side reactions. (I) Reduction of the ring-fission product. (II) Conversion of benzoate to *p*-hydroxybenzoate. (III) Mono- and di- hydroxybiphenyls.

been recorded (Bestetti et al. 1989). The growth substrate was catabolised via 2-phenyl-2-propen-1-ol and 1,2-dihydroxy-3-isopropenyl-3-cyclohexene, implying a pathway different to that previously reported (Omori et al. 1974). Furthermore, the strain was also able to biotransform styrene to 1,2-dihydroxy-3-ethenyl-3-cyclohexene.

There have been several tentative reports of the involvement of *meta*-cleavage of alkenylbenzenes. Sielicki et al. (1978) observed the development of yellow culture fluids during stationary phase of growth of a mixed culture with styrene as the sole source of carbon and energy. (The appearance of yellow culture fluids often occurs during the degradation of aromatic compounds caused by the accumulation *meta*-cleavage products.) This observation was not further discussed. Hartmans et al. (1989) observed a transient accumulation of a yellow product during growth of *Xanthobacter* strain 124X on styrene and 1-phenylethanol. They reported that the compound had different spectral properties to those reported for the ring-cleavage product of the catechol of 1-phenylethanol (2,7-dihydroxy-6-oxoocta-2,4-dienoate – Cripps et al. 1978) but did not identify the product. In contrast, Dzhusupova et al. (1985) showed induced levels of protocatechuate 3,4-dioxygenase (an intra-diol cleaving enzyme) in strains of *Pseudomonas* grown on α-methylstyrene, suggesting a novel pathway, sadly without explanation.

Clearly from the foregoing discussion alkenylbenzenes are degraded by bacteria and it seems certain, from the circumstantial evidence, that ring-cleavage must occur.

Biphenyl

Biphenyl may be considered as a substituted benzene even though the substituent is in fact benzene itself; biphenyl is not a polycyclic aromatic hydrocarbon. The catabolism of biphenyl has received much attention recently due to the increasing concern over the fate of polychlorinated (PCBs) which are established as worldwide pollutants. Previous review articles have discussed the biodegradation

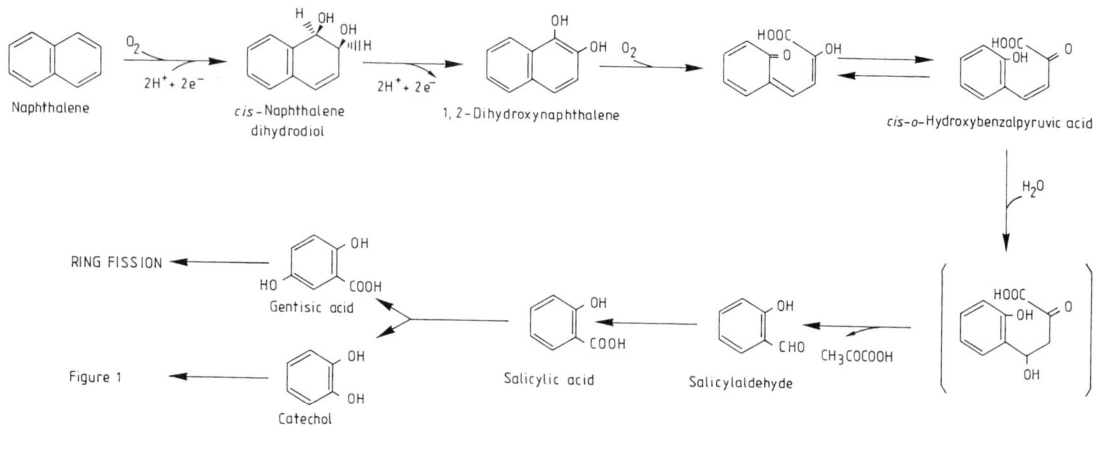

Fig. 11. The initial reactions of naphthalene biodegradation.

of biphenyl. Some of these have highlighted two key pathways for the catabolism of biphenyl (Cripps & Watkinson 1978, for exampe). Recently we (Smith & Ratledge 1989b) have re-investigated the so-called alternative pathway (Lunt & Evans 1970) and demonstrated it to be erroneous (see below).

The pathway for the biodegradation of biphenyl is given in Fig. 10. Initial attack of biphenyl proceeds via 2,3-dihydro-2,3-dihydroxybiphenyl (Catelani et al. 1971; Catelani et al. 1973; Gibson et al. 1973) and 2,3-dihydroxybiphenyl (Catelani et al. 1973; Smith & Ratledge 1989b). The catechol type compound is then ring cleaved between carbon atoms 1 and 2 to form 2-hydroxy-6-oxo-6-phenylhexa-2,4-dienoate (Catelani et al. 1973; Catelani & Colombi 1974; Ishigooka et al. 1986; Smith & Ratledge 1989b). Further catabolism of this ring fission-product is via 2-oxo-penta-4-enoate and benzoate (Smith & Ratledge 1989b). This pathway is common to many species of bacteria.

Contemporary investigators have revealed some subtle differences between different bacteria. We found that benzoate was a dead-end product in the biodegradation by *Pseudomonas* sp. NCIB 10643, whereas *Nocardia* sp. NCIB 10503 catabolised the benzoate via oxidative-decarboxylation to catechol which was subsequently degraded via the β-ketoadipate pathway (Smith & Ratledge 1989b). Omori et al. (1986) demonstrated the NADPH dependent reduction of the ring-cleavage product to 2,6-dioxo-6-phenylhexanoic acid in biphenyl grown cells of *Pseudomonas cruciviae*, although no physiological significance of this enzyme step was postulated. 3-Hydroxybenzoate and cinnamic acid were also reported as intermediates in the plasmid encoded catabolism of biphenyl by a strain of *Pseudomonas putida* (Starovoitov et al. 1986). No scheme of dissimilation was proposed to account for the production of cinnamic acid. Furukawa & Suzuki (1988) transferred the plasmid encoding for the degradation of biphenyl in *Pseudomonas pseudoalcaligenes* into *Pseudomonas aeruginosa* and subsequently detected 2,3,2′,3′-tetrahydroxybiphenyl, suggesting that the initial two enzymes of biphenyl degradation had a broader substrate specificity than hitherto considered. Khan & Walia (1990) have recently cloned the genes encoding for two of the key enzymes of biphenyl biodegradation (3-phenylcatechol dioxygenase and 2-hydroxy-6-oxo-6-phenylhexa-2,4-dienoate hydrolase) from *Pseudomonas putida* into *Escherichia coli*.

Our own investigation with *Nocardia* sp. NCIB 10503 (Smith & Ratledge 1989b) was the first report of the complete degradation of biphenyl by an actinomycete. Previously, Schwartz (1981) reported growth of such a bacterium but only identified mono-hydroxybiphenyls and 2,2′-dihydroxybiphenyl as intermediates (Fig. 10). It remains to be seen if other members of the actinomycetes biodegrade biphenyl via mono-oxygenase type enzymes rather than the more conventional dioxygenases.

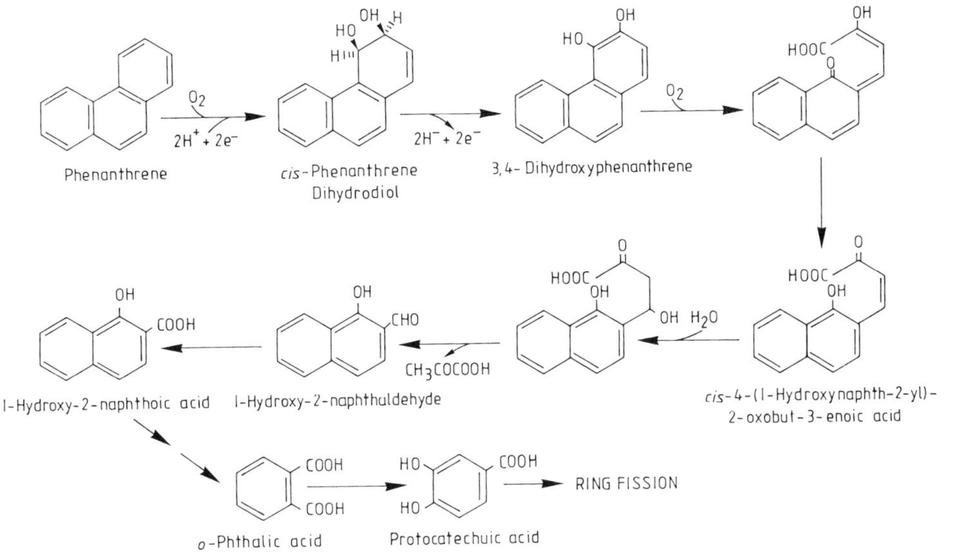

Fig. 12. Various proposed steps in the biodegradation of phenanthrene.

Fused-ring aromatic compounds

Naphthalene

Analogous with benzene, the bacterial degradation of naphthalene has frequently been reported over the last twenty years. The biodegradative route employed by the vast majority of micro-organisms is given in Fig. 11 and evidence to support this pathway can be found in previous review articles (Cerniglia 1984; Gibson & Subramanian 1984). In common with most other types of bacterial biodegradation of aromatic compounds, the initial attack proceeds via the action of dioxygenase attack. The resultant dihydrodiol is then converted to 1,2-dihydroxynaphthalene by dehydrogenase type enzymes. This catechol type compound then undergoes extradiol type cleavage between carbon atoms 1 and 9. The catabolic divergence in the catabolism of salicylate (see Fig. 11), may be somewhat misleading as it has only been reported in *Pseudomonas fluorescens*; the norm being the oxidative decarboxylation of the salicylate to yield catechol (Gibson & Subramanian 1984).

The catabolism of naphthalene by pseudomonads is often plasmid encoded as originally reported by Dunn & Gunsalus (1973). Subsequent studies have demonstrated that the genes are located on two operons (*nah* and *sal*) cited on plasmids of about 80-kilobase-pairs. The *nah* operon encodes for the genes for the conversion of naphthalene to salicylate, whilst the *sal* operon encodes for the conversion of salicylate to central metabolites (You et al. 1988). Recent reports on the genetic basis of naphthalene biodegradation have dissected the structure and function of the individual genes (Schell 1986; Schell & Wender 1986; You & Gunsalus 1986; You et al. 1988) but this work falls outside the scope of this present review.

The association and uptake of naphthalene in *Pseudomonas putida* was recently reported (Bateman et al. 1986). Neither an energised membrane or ATP were essential for the association or uptake of naphthalene. This association was not influenced by the catabolic plasmid.

Durham & Stewart (1987) proposed the recruitment of naphthalene dissimilatory enzymes of *Pseudomonas putida* for the oxidation of 1,4-dichloronaphthalene to 3,6-dichlorosalicylate (a precursor for the synthesis of the herbicide Dicamba, 3,6-dichloro-2-methoxybenzoate). However, the reported conversion was very low, being only 1% of the rate of salicylate conversion.

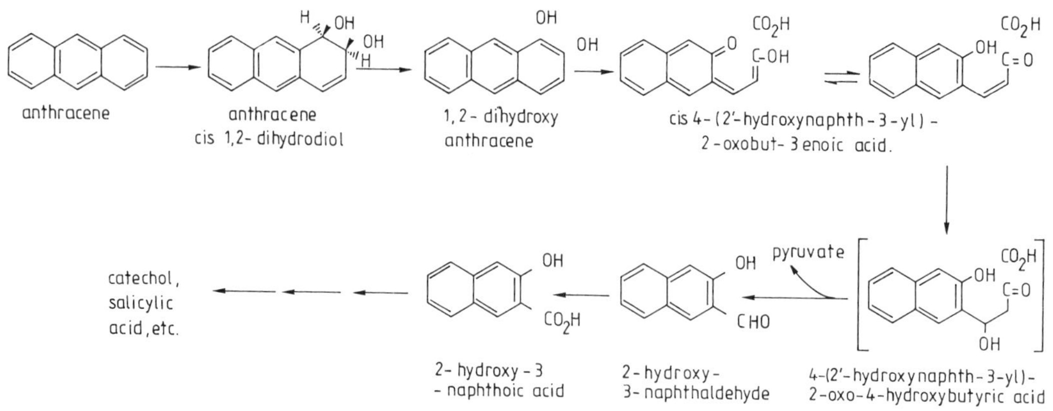

Fig. 13. Various proposed steps in the biodegradation of anthracene.

Polycyclic aromatic hydrocarbons

There is currently great concern about the levels of polycyclic hydrocarbons in the environment. The presence in the environment of large quantities may be attributed to both petrogenic and pyrogenic sources (Laflamme & Hite 1978) and these compounds are considered extremely undesirable as most are potential carcinogens, mutagens and tetragens. There has consequently been much research performed into the biodegradation of many these compounds as this is the major route through which polycyclic aromatic compounds are dissimilated. The biodegradative routes for phenanthrene and anthracene have been propsed (Figs 12 and 13, respectively) and the reader is directed to previous review articles (Gibson & Subramanian 1984; Cerniglia & Heitkamp 1989). There remain many unanswered questions regarding the biodegradation of these compounds. To date only the initial attack of these compounds has been reported and there have been no detailed investigations into the enzymes involved or the modes of ring-fission.

Foght et al. (1990) recently reported on the complete biodegradation of [^{14}C]phenanthrene in mineral oil by various different bacteria; the use of radio-labelled substrates offers conclusive evidence that the compound is biodegraded.

The biodegradation of higher molecular weight polycyclic aromatic compounds is even less well understood. There have been a few reports on the co-metabolism of these compounds (see Cerniglia & Heitkamp 1989). Until recently the only reports concerned the initial oxidation of high molecular weight polycyclic compounds by bacteria grown on alternative aromatic compounds and reflected the substrate specificity of the enzymes rather than novel modes of biodegradation. However, there have been several recent advances in this area. Heitkamp & Cerniglia (1988) isolated a single Gram-positive strain capable of the biodegradation of naphthalene, phenanthrene, fluoranthene and pyrene. The complete biodegradation of fluoranthene by pure cultures of *Pseudomonas paucimobilis* was demonstrated by Mueller et al. (1990). This was partially attributable to the use of Tween 80 in the medium to increase the bioavailability of the substrate. Fluoranthene grown cells were able to oxidise a wide range of aromatic compounds (anthraquinone, benzo[*b*]fluorene, biphenyl, chrysene and pyrene. These authors suggested that this was due to novel biodegradative routes but no route(s) have yet been postulated. Weissenfels et al. (1990) isolated various pure cultures capable of biodegrading different polycyclic aromatic compounds; *Pseudomonas paucimobilis* which grew on phenanthrene, *Pseudomonas vesicularis* capable of degrading fluorene and *Alcaligenes denitrificans* which grew on fluoranthene. No details of the biochemical pathways were included.

Concluding remarks

Although the literature abounds with reports on the bacterial biodegradation of aromatic compounds (and from this one can deduce that there exists a wealth of knowledge in this area) there do still exist gaps in our understanding.

The recent discovery of pure bacterial strains capable of anaerobic growth on toluene (Lovley & Lonergan 1990) raises the intriguing question: are there organisms that grow anaerobically on other aromatic hydrocarbons?

As outlined above, bacterial strains that degrade *p*- and *m*-xylenes cannot attack *o*-xylene (and vice versa) and this presents interesting questions regarding the evolution of these strains. It is also curious that there have only been two reports of the degradation of the *o*-isomer (Baggi et al. 1987; Schraa et al. 1987). There have been very few reports on the biodegradation of other di- and polyalkylsubstituted benzenes and this remains an area open to future studies.

With regard to alkenylbenzenes, the entire topic requires in depth studies to elucidate the pathways leading to ring-fission, the nature of the ring-fission mechanisms and the subsequent catabolism to central metabolites.

The increasing awareness of the occurrence of polycyclic aromatic hydrocarbons in the environment has provoked great research efforts into their biodegradation. The number and complexity of compounds now known to dissimilated by pure bacterial strains is rapidly increasing. As the problems, such as water solubility, are overcome, so the list is expected to grow. It is also hoped that the promising reports on the isolation of bacteria capable of growing on these compounds are followed up by detailed studies into the exact nature of the pathways employed.

An area of aromatic hydrocarbon biodegradation not covered in this review is that of mixed substrate interactions. In the environment it is unlikely that the micro-organisms are faced with single aromatic hydrocarbons; mixtures are much more likely to occur. It is well established that certain mixtures are more rapidly biodegraded than when the compounds are present singularly (McCarty et al. 1984) however, these studies are mainly based on work with non-growth-supporting (secondary) substrates, with biomass being created by one or more easily degraded primary substrate, present in high concentrations. Little is known about substrate interactions among biodegradable aromatic hydrocarbons present in growth supporting concentrations. Recently, Bauer & Capone (1988) investigated the biodegradation of mixtures of polycyclic aromatic hydrocarbons. Amongst their findings they showed that naphthalene stimulated the biodegradation of phenanthrene but not that of anthracene. Arvin et al. (1989) demonstrated the interaction of aromatic substrates during the biodegradation of benzene. The presence of toluene or xylene stimulated the degradation of benzene but that toluene and xylene had an antagonistic effect on the utilisation benzene. We (Smith et al. 1991) observed complete inhibition of growth of *Pseudomonas* sp. when presented with a mixture of biphenyl and ethylbenzene; both compounds being readily degraded when present singularly. None of these groups could explain their data and further research is required to establish if this is a commonly occurring phenomena and to elucidate the mechanisms involved.

It is possible that the answer to the above observations lies in the mechanisms of association and uptake of aromatic hydrocarbons by bacteria. This is an area of research which has been much neglected and intensive studies are needed to clarify this.

References

Amund OO & Higgins IJ (1985) The degradation of 1-phenylalkanes by an oil degrading strain of *Acinetobacter lwoffi*. Antonie van Leeuwenhoek 51: 45–56

Arvin E, Jensen BK & Gundersen AT (1989) Substrate interactions during the areobic degradation of benzene. Appl. Environ. Microbiol. 55: 3221–3225

Axell BC & Geary PJ (1975) The metabolism of benzene by bacteria. Biochem. J. 136: 927–934

Baggi G, Catelani D, Galli E & Treccani V (1972) The microbial degradation of phenylalkanes. Biochem. J. 126: 1091–1097

Baggi G, Boga MM, Catelani D, Galli E & Trecccani V (1983) Styrene catabolism by a strain of *Pseudomonas fluorescens*. Syst. Appl Microbiol. 4: 141–147

Baggi G, Barbieri P, Galli E & Tollari S (1987) Isolation of a

Pseudomonas stutzeri strain that degrades *o*-xylene. Appl. Envriron. Microbiol. 53: 2129–2132

Ballard DGH, Courtis A, Shirley IM & Taylor SC (1983) A biotech route to polyphenylene. J. Chem. Soc., Chem. Comm. pp 954–955

Bateman JN, Speers B, Feduik L & Hartline RA (1986) Naphthalene association and uptake in Pseudomonas putida. J. Bacteriol. 166: 155–161

Bauer JE & Capone DG (1988) Effects of co-occurring aromatic hydrocarbons on the degradation of individual polycyclic aromatic hydrocarbons in marine sediment slurries. Appl. Environ. Microbiol. 54: 1649–1655

Bayly RC & Barbour MG (1984) The degradation of aromatic compounds by the *meta* and gentisate pathways. In: Gibson DT (Ed) Microbial Degradation of Organic Compounds (pp 253–294). Marcel Dekker, New York

Berry DF, Francis AJ & Bollag J-M (1987) Microbial metabolism of homocyclic aromatic compounds under anaerobic conditions. Microbiol. Revs. 51: 43–49

Bestetti G & Galli E (1984) Plasmid-coded degradation of ethylbenzene and 1-phenylethanol in *Pseudomonas fluoresens*. FEMS Microbiol. Letts. 21: 165–168

Bestetti G, Galli E, Benigini C Orsini F & Pelizzoni F (1989) Biotransformation of styrenes by a *Pseudomonas putida*. Appl. Microbiol. Biotechnol. 30: 252–256

Burlage RS, Hooper SW & Sayler GS (1989) The TOL (pWWO) catabolic plasmid. Appl. Environ. Microbiol 55: 1323–1328

Catelani D, Sorlini C & Treccani V (1971) The metabolism of biphenyl by *Pseudomonas putida*. Experientia 27: 1173–1174

Catelani D, Colombi A, Sorlini C & Treccani V (1973) Metabolism of biphenyl. 2-Hydroxy-6-oxo-6-phenylhexa-2,4-dienoate: the *meta* cleavage product from 2,3-dihydroxybiphenyl by *Pseudomonas putida*. Biochem. J. 134: 1063–1066

Catelani D & Colombi A (1974) Metabolism of biphenyl. Structure and physical properties of 2-hydroxy-6-oxo-6-phenylhexa-2,4-dienoate, the *meta* cleavage product from 2,3-dihydroxybiphenyl by *Pseudomonas putida*. Biochem. J. 143: 431–434

Catelani D, Colombi A, Sorlini C & Treccani V (1977) Metabolism of quaternary carbon compounds: 2,2-dimethylheptane and tertbutylbenzene. Appl. Environ. Microbiol. 34: 351–354

Cerniglia CE (1984) Microbial metabolism of polycyclic aromatic compounds. Adv. Appl. Microbiol. 30: 31–71

Cerniglia CE & Heitkamp MA (1989) Microbial degradation of polycyclic compounds (PAH) in the aquatic environment In: Varanasi V (Ed) Metabolism of PAHS in the Aquatic Environment (pp 41–68). CRC Press Inc. Boca Raton, Florida

Cripps RE, Trudgill PW & Wheatley JG (1978) The metabolism of 1-phenylethanol and actophenone by *Nocardia* T5 and an *Arthrobacter* species. Eur. J. Biochem. 86: 175–186

Cripps RE & Watkinson RJ (1978) Polycyclic aromatic hydrocarbons: Metabolism and environmental aspects. In: Watkinson RJ (Ed) Developments in the Biodegradation of Hydrocarbons (pp 113–134). Applied Science Publishers, London

Dagley S (1981) New perspectives in aromatic catabolism. In: Leisinger T, Cook AM, Hütter R & Nüesch J (Eds) Microbial Degradation of Xenobiotics and Recalcitrant Compounds (pp 181–186). Academic Press, New York

Dagley S (1986) Biochemistry of aromatic hydrocarbon degradation in Pseudomonads. In: Sokatch JR (Ed) The Bacteria (Vol 10) (pp 527–555). Academic Press, New York

Davey JF & Gibson DT (1974) Bacterial metabolism of *p*- and *m*-xylene: Oxidation of the methyl substituent. J. Bacteriol. 119: 923–929

Davis RS, Hossler, FE & Stone RW (1968) Metabolism of *p*- and *m*-xylene by species of *Pseudomonas*. Can. J. Microbiol 27: 1005–1009

DeFrank JJ & Ribbons DW (1976) The *p*-cymene pathway in *Pseudomonas putida* PL: Isolation of a dihydrodiol accumulated by a mutant. Biochem. Biophys. Res. Commun. 70: 1129–1135

DeFrank JJ & Ribbons DW (1977a) *p*-Cymene pathway in *Pseudomonas putida*: Initial reactions. J. Bacteriol. 129: 1356–1364

DeFrank JJ & Ribbons DW (1977b) *p*-Cymene pathway in Pseudomonas putida: ring cleavage of 2,3-dihydroxy-p-cumate and subsequent reactions. J. Bacteriol. 129: 1365–1375

Duggleby CJ & Williams PA (1986) Purification and some properties of the 2-hydroxy-6-oxohepta-2,4-dienoate hydrolase (2-hydroxymuconic semialdehyde hydrolase) encoded by the TOL plasmid pWWO from *Pseudomonas putida* mt 2. J Gen Microbiol 132: 717–726

Dunn NW & Gunsalus IC (1973) Transmissible plasmid coding for the early enzymes of naphthalene oxidation in *Pseudomonas putida*. J. Bacteriol. 114: 974–979

Durham DR & Stewart DB (1987) Recruitment of naphthalene dissimilatory enzymes for the oxidation of 1,4-dichloronaphthalene to 3,6-dichlorosalicylate, a precursor of the herbicide Dicamba. J. Bacteriol. 169: 2889–2892

Dzhusupova DB, Baskunov BP, Golovleva LA, Alieva RM & Ilyaletdinov AN (1985) Peculiarities of the oxidation of α-methylstyrene by bacteria of the genus *Pseudomonas*. Mikrobiologiya 54: 136–140

Eaton RW & Timmis KN (1986) Characterization of a plasmid-specified pathway for catabolism of isopropylbenzene in *Pseudomonas putida* RE 204. J Bacteriol 168: 123–131

Evans WC & Fuchs G (1988) Anaerobic degradation of aromatic compounds. Annu. Rev. Microbiol 42: 289–317

Foght JM, Fedorak PM & Westlake DWS (1990) Mineralization of [^{14}C]hexadecane and [^{14}C]phenanthrene in crude oil: Specificity amoung bacterial isolates. Can. J. Microbiol 36: 169–175

Franklin FCH, Bagdasarian M, Bagdasarian MM & Timmis KN (1981) Molecular and functional analysis of TOL plasmid pWWO from *Pseudomonas putida* and cloning of the entire regulated aromatic ring meta cleavage pathway. Proc. Acad. Sci. USA 78: 7458–7462

Furukawa K & Suzuki H (1988) Gene manipulation of catabolic activities for production of intermediates of various biphenyl compounds. Appl. Microbiol. Biotechnol. 29:363–369

Gibson DT (1971) The microbial oxidation of aromatic compounds. Crit. Rev. Microbiol. 1:199–223

Gibson DT, Koch JR & Kallio RE (1968) Oxidative degradation of aromatic hydrocarbons by micro-organisms.I Enzymatic formation of catechol from benzene. Biochemistry 7:2643–2656

Gibson DT, Cardini GE, Maesels FC & Kallio RE (1970) Incorporation of ^{18}O into benzene. Biochemistry 9:1631–1635

Gibson DT, Roberts RL, Wells MC & Kobal VM (1973) Oxidation of biphenyl by a *Beijerinckia* species. Biochem. Biophys. Res. Commun. 50:211–219

Gibson DT, Gschwendt B, Yeh WK & Kobal VM (1973) Initial reaction in the oxidation of ethylbenzene by *Pseudomonas putida*. Biochemistry12:1520–1528

Gibson DT, Mahadevan V & Davey JF (1974) Bacterial metabolism of *p*- and *m*-xylene: Oxidation of the aromatic ring. J. Bacteriol. 119: 930–936

Gibson DT & Subramanian V (1984) Microbial Degradation of aromatic hydrocarbons. In: Gibson DT (Ed) Microbial Degradation of Organic Compounds (pp 361–369). Marcel Dekker New York

Hartmans S, Smits, JP, van der Werf MJ, Volkering F & de Bont JAM (1989) Metabolism of styrene oxide ande 2-phenylethanol in the styrene-degrading *Xanthobacter* strain 124X. Appl. Environ. Microbiol. 55: 2850–2855

Hartmans S, van der Werf MJ & de Bont JAM (1990) Bacterial degradation of styrene involving a novel flavin adenine dinucleotide-dependent styrene monooxygenase. Appl. Environ. Microbiol. 56: 1347–1351

Heitkamp MA & Cerniglia CE (1988) Mineralization of polycyclic aromatic hydrocarbons by a bacterium isolated from sediment below an oil field. Appl. Environ. Microbiol. 54: 1612–1614

Högn T & Jaenicke L (1972) Benzene metabolism of *Moraxella* sp. Eur. J. Biochem. 30: 369–375

Hooper DJ (1978) Microbial degradation of aromatic hydrocarbons. In: Watkinson RJ (Ed) Developments in the Biodegradation of Hydrocarbons (pp 85–112). Applied Science Publishers, London

Ishigooka H, Yashida Y, Omori T & Minoda Y (1986) Enzymatic dioxygenation of biphenyl 2,3-diol and 3-isopropylcatechol. Agric. Biol. Chem. 50: 1045–1046

Jigami Y, Omori T & Minoda Y (1974a) A new ring fission product suggestive of a unknown reductive step in the degradation of n-butylbenzene by *Pseudomonas*. Agric. Biol. Chem. 38: 1757–1759

Jigami Y, Omori T, Minoda Y & Yamada K (1974b) Formation of of 3-ethylsalicylic acid from 3-ethyltoluene by *Pseudomonas ovalis*. Agric. Biol. Chem. 38: 467–469

Jigami Y, Omori T & Minoda Y (1975) The degradation of isopropylbenzene and isobutylbenzene by *Pseudomonas* sp. Agric. Biol. Chem. 39: 1817–1788

Khan AA & Walia SK (1990) Identification and localization of 3-phenylcatechol dioxygenase and 2-hydroxy-6-oxo-6-phenylhexa-2,4-dienoate hydrolase genes of *Pseudomonas putida* and expression in *Eschericha coli*. Appl. Environ. Microbiol. 56: 956–962

Kunz DA & Chapman PJ (1981) Catabolism of pseudocumene and 3-ethyltoluene by *Pseudomonas putida* (*arvilla*) mt-2: evidence for new functions of the TOL (pWWO) plasmid. J. Bacteriol. 146: 179–191

Laflamme RE & Hite RA (1978) The global distribution of polycyclic aromatic hydrocarbons in recent sediments. Geochim. Cosmomchim. Acta. 42: 289–303

Ley SV, Sternfeld & Taylor SC (1987) Microbiol oxidation in synthesis: a six step preparation of (+)-pinitol from benzene. Tetrahydron Lett. 28: 225–226

Lovley DR & Lonergan DJ (1990) Anaerobic oxidation of toluene, phenol and *p*-cresol by the dissimilatory iron-reducing organism, GS-15. Appl. Environ Microbiol. 56: 1858–1864

Lunt DO & Evans WC (1970) The microbial metabolism of biphenyl. Biochem J 118: 54P

Marr EK & Stone RW (1961) Bacterial oxidation of benzene. J. Bacteriol. 85: 425–430

McCarty PL, Rittmann BE & Bouwer EJ (1984) Microbial processes affecting chemical transformations in groundwater. In: Bitton G & Gerba CP (Eds) Groundwater Pollution Micobiology (pp 89–115). John Wiley & Sons, New York

Mueller JG, Chapman PJ, Blattmann BO & Pritchard PH (1990) Isolation and characterization of a fluoranthene-utilizing strain of *Pseudomonas paucimobilis*. Appl Environ. Microbiol. 56: 1079–1086

Nakazawa T, Inouye S & Nakazawa A (1980) Physical and functional analysis of RP4-TOL plasmid recombinants: Analysis of insertion and deletion mutants. J. Bacteriol. 144: 222–231

Omori T, Jigami Y & Minoda Y (1974) Microbial oxidation of α-methylstyrene and β-methylstyrene. Agric. Biol. Chem. 38: 409–415

Omori T, Ishigooka H & Minoda Y (1986) Purification and some properties of 2-hydroxy-6-oxo-6-phenylhexa-2,4-dienoic acid (HODPA) reducing enzyme from *Pseudomonas cruciviae* S93B1, involved in the degradation of biphenyl. Agric. Biol. Chem. 50: 1513–1518

Ramos JL, Mermod N & Timmis KN (1987) Regulatory circuits controlling transcription of TOL plasmid operon encoding *meta*-cleavage pathway for degradation of alkylbenzoates by *Pseudomonas*. Mol. Microbiol. 1: 293–300

Sala-Trepat JM, Murray K & Williams PA (1972) The metabolic divergence in the *meta* cleavage pathway of catechols by *Pseudomonas putida* NCIB 10015. Eur. J. Biochem. 28: 347–356

Sariaslani FS, Harper DB & Higgins IJ (1974) Microbial degradation of hydrocarbons. Biochem. J. 140: 31–45

Schell MA (1986) Homology between nucleotide sequences of promoter regions of *nah* and *sal* operons of NAH7 plasmid of *Pseudomonas putida*. Proc. Natl. Acad. Sci. USA 83: 369–373

Schell MA & Wender PE (1986) Indentification of the *nahR* genee product and nucleotide sequence required for its activation of the *sal* operon. J. Bacteriol. 166: 9–14

Schraa G, Bethe BM, van Neerven ARW, van den Tweel WJJ, van der Wende E & Zehnder AJB (1987) Degradation 1,2-dimethylbenzene by *Corynebacterium* strain C125. Antonie van Leewenhoek 53: 159–170

Schwartz RD (1981) A novel reaction: *meta* hydroxylation of biphenyl by an actinomycete. Enzyme Microbial Technol. 3: 158–159

Shirai K (1986) Screening microorganisms for catechol production from benzene. Agric. Biol. Chem. 50: 2875–2880

Shirai K (1987) Catechol production from benzene through reaction with resting and immobilized cells of a mutant strain of *Pseudomonas*. Agric. Biol. Chem. 51: 121–128

Shirai K & Hisatsuka K (1979) Isolation and identification of styrene assimiliating bacteria. Agric. Biol. Chem. 43: 1595–1596

Simpson HD, Green J & Dalton H (1987) Purification and some properties of a novel heat-stable *cis*-toluene dihydrodiol dehydrogenase. Biochem. J. 244: 585–590

Sielicki M, Focht DD & Martin JP (1978) Microbial transformations of styrene and [^{14}C]styrene in soil and enrichment cultures. Appl. Environ. Microbiol. 35: 124–128

Smith MR & Ratledge C (1989a) Catabolism of alkylbenzenes by *Pseudomonas* sp. NCIB 10643. Appl. Microbiol. Biotechnol. 32: 68–75

Smith MR & Ratledge C (1989b) Catabolism of biphenyl by *Pseudomonas* sp. NCIB 10643 and *Nocardia* sp. NCIB 10503. Appl. Microbiol. Biotechnol. 30: 395–401

Smith MR, Ewing M & Ratledge C (1991) The interactions of various aromatic substrates degraded by *Pseudomonas* sp. NCIB 10643: Synergistic inhibition of growth by two compounds which serve as growth substrates. Appl. Micobiol. Biotechnol. (in press)

Starovoitov II, Selfionov SA, Nefedova MY & Adanin VM (1986) Catabolism of diphenyl by *Pseudomonas putida* strain BS893 containing biodegradation plasmid pBS241. Microbiology (Engl. Transl.) 54: 726–727

van den Tweel WJJ, Janssens & de Bont JAM (1986) Degradation of 4-hydroxyphenylacetate by *Xanthobacter* 124X. Antonie van Leeuwenhoek 52: 309–318

van den Tweel WJJ, Vorage MJAW, Marsman EH, Koppejan J, Tramper J & de Bont JAM (1988) Continuous production of *cis*-1,2-dihydroxycyclohexa-3,5-diene (*cis*-benzeneglycol) from benzene by a mutant of a benzene degrading *Pseudomonas* sp. Enzyme Microbial Technol. 10: 134–142

Vecht SE, Platt MW, Er-El Z & Goldberg I (1988) The growth of *Pseudomonas putida* on *m*-toluic acid and on toluene in batch and chemostat cultures. Appl. Microbiol. Biotechnol. 27: 587–592

Weissenfels WD, Beyer M & Klein J (1990) Degradation of fluorene and fluoranthene by pure bacterial cultures. Appl. Microbiol. Biotechnol. 32: 479–484

Wigmore GJ, Bayley RC & DiBerardino D (1974) *Pseudomonas putida* mutants defective in the catabolism of the products of meta fission of catechol and its methyl analogues. J. Bacteriol. 120: 31–37

Winstanley C, Taylor SC & Williams PA (1987) pWW174: A large plasmid from *Acinetobacter calcoaceticus* encoding benzene catabolism by the β-ketoadipate pathway. Mol. Microbiol. 1: 219–227

You IS & Gunsalus IC (1986) Regulation of the nah and sal operons of plasmid NAH7: Evidence for a new function in nahR. Biochem. Biophys. Res. Commun. 141: 986–992

You IS, Ghosal D & Gunsalus IC (1988) Nucleotide sequence of plasmid NAH7 *gene nahR* and DNA binding of the *nahR* product. J. Bacteriol. 170: 5409–5415

Zeyer J, Eicher P, Dolfing J & Schwarzenbach RP (1990) Anaerobic degradation of aromatic compounds. In: Kamely D, Chakrabarty A & Omenn GS (Eds) Biotechnology and Biodegradation (pp 33–40). Gulf Publishing Company, Houston

Biodegradation **1**: 207–220, 1990.

Degradation of halogenated aromatic compounds

L.C.M. Commandeur & J.R. Parsons*
Department of Environmental and Toxicological Chemistry, University of Amsterdam, Nieuwe Achtergracht 166, 1018 WV Amsterdam, The Netherlands

Key words: halobenzoates, halobenzenes, halophenols, haloanilines, halophenoxyacetates, halobiphenyls, halodibenzo-*p*-dioxins, halodibenzofurans, hydrolytic dehalogenation, oxidative dehalogenation, reductive dehalogenation

Abstract

Due to their persistence, haloaromatics are compounds of environmental concern. Aerobically, bacteria degrade these compounds by mono- or dioxygenation of the aromatic ring. The common intermediate of these reactions is (halo)catechol. Halocatechol is cleaved either intradiol (*ortho*-cleavage) or extradiol (*meta*-cleavage). In contrast to *ortho*-cleavage, *meta*-cleavage of halocatechols yields toxic metabolites. Dehalogenation may occur fortuitously during oxygenation. Specific dehalogenation of aromatic compounds is performed by hydroxylases, in which the halo-substituent is replaced by a hydroxyl group. During reductive dehalogenation, haloaromatic compounds may act as electron-acceptors. Herewith, the halo-substituent is replaced by a hydrogen atom.

Abbreviations: CBz – chlorobenzene, DCBz – dichlorobenzene, TrCBz – trichlorobenzene, TCBz – tetrachlorobenzene, PCBz – pentachlorobenzene, HCBz – hexachlorobenzene, CBA – chlorobenzoic acid, BBA – bromobenzoic acid FBA – fluorobenzoic acid, IBA – iodobenzoic acid, CP – chlorophenol, CA – chloroaniline, PCBs – polychlorinated biphenyls, CB – chlorobiphenyl, 2,4-D – 2,4-dichlorophenoxyacetic acid, 2,4,5-T – 2,4,5-trichlorophenoxyacetic acid

Introduction

Halogenated aromatic compounds have been produced industrially on a large scale for several decades. Such chemicals, particularly the chlorinated ones, have been widely used as pesticides (e.g. DDT, 2,4-D, 2,4,5-T, chlorophenols) or for other industrial uses (e.g. PCBs in electrical equipment and as hydraulic fluids). Others, such as PCDDs and PCDFs are produced unintentionally as trace contaminants during industrial syntheses and incinerations. Brominated aromatic compounds have found use as flame retardants. Fluorinated and iodinated aromatic compounds are components of pharmaceutical agents. The chemical inertness and hydrophobicity of many of these compounds has resulted in them becoming widely distributed in the environment; in particular accumulating in many terrestial and aquatic organisms. This, coupled with their toxicity, has given rise to concern about their fate in the environment.

Despite the fact that naturally occurring halogenated aromatic compounds are rare, many bacteria have been isolated which can degrade such chemicals. These bacteria are often unable to grow on these compounds, but are able to degrade them while growing on other compounds, such as their nonhalogenated analogues. This process is referred to as co-metabolism.

The pathways by which halogenated aromatic

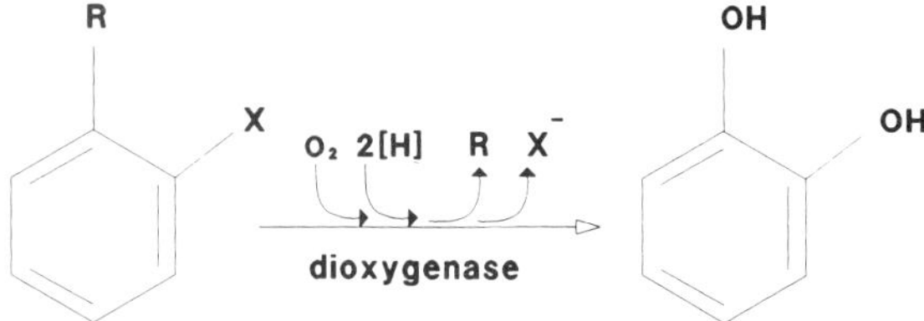

Fig. 1. Oxidative dehalogenation of haloaromatic compounds. R = e.g. COOH, H, NH_2. X = F, Cl, Br, I.

compounds are degraded by microorganisms are similar to those for the degradation of aromatic compounds in general. Under aerobic conditions aromatic compounds are transformed by mono- and di-oxygenation into dihydroxylated derivatives before ring cleavage takes place. Under anaerobic conditions, degradation follows reductive pathways. The aromaticity is lost before ring cleavage. The degradative pathways for aromatic compounds are described in detail elsewhere in this issue by Smith.

Dehalogenation of aromatic compounds

As halogen substituents of halogenated aromatic compounds are, to a large extent, responsible for their properties, removal of this substituents is a key step in their degradation. In many cases, dehalogenation of aromatic compounds occurs after the ring system is cleaved. Examples are the dehalogenation of halocatechols during ring cleavage by lactonization (Fig. 4). However, direct dehalogenation of aromatic compounds, without loss of aromaticity has been demonstrated. Three forms of such reactions are known:

1. Oxidative dehalogenation in which the halogen is lost fortuitously during oxygenation of the ring (Fig. 1). This reaction occurs only under aerobic conditions.
2. Hydrolytic dehalogenation in which a halogen (for example at the para position) is specifically replaced by a hydroxyl group (Fig. 2). The oxygen atom in the hydroxyl group is derived from water instead of from oxygen. This reaction can

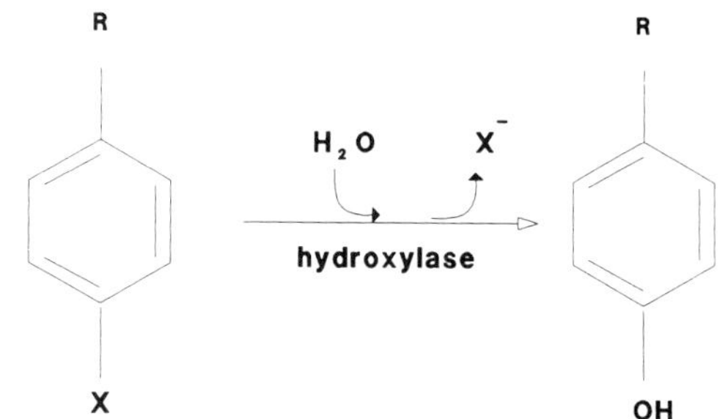

Fig. 2. Hydrolytic dehalogenation of haloaromatic compounds. R = e.g. COOH, OH, NH_2. X = F, Cl, Br, I.

occur under both aerobic and denitrifying conditions.
3. Reductive dehalogenation in which the halogen is replaced by a hydrogen (Fig. 3). This reaction occurs almost exclusively under sulfogenic and methanogenic conditions. It has been proposed that the halogenated aromatic compound acts as a terminal electron acceptor. In one case it has been shown that reductive dehalogenation is coupled to growth and to ATP formation (see below) and thus can be referred to respiration.

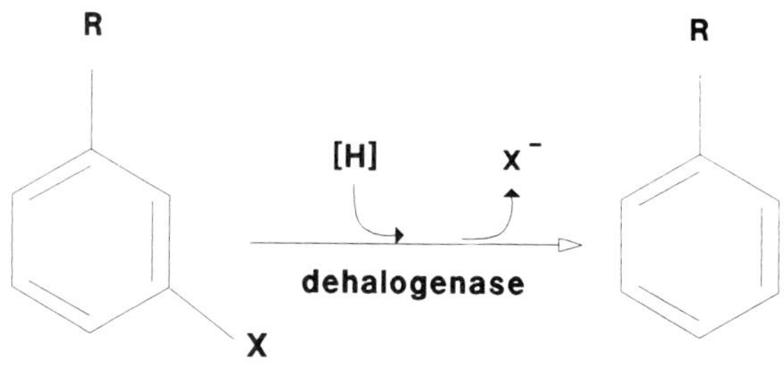

Fig. 3. Reductive dehalogenation of haloaromatic compounds. R = COOH, H, OH, NH_2, C_6H_5. X = F, Cl, Br, I.

Degradation of halogenated benzoic acids

Halogenated benzoates have been used extensively as model compounds to study the degradation of haloaromatics. The degradation of halogenated benzoate was reviewed earlier by Reineke (1984) and recently by Reineke & Knackmuss (1988).

Aerobically, halogenated benzoates are mainly degraded by dioxygenation of the aromatic ring

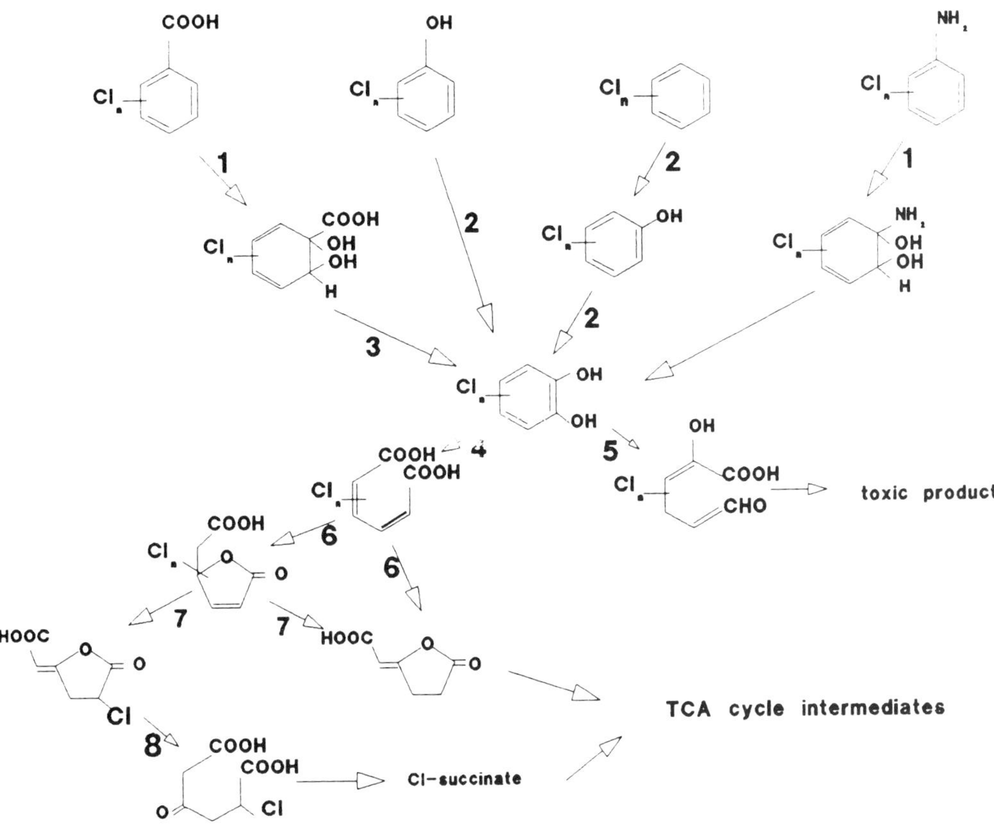

Fig. 4. Metabolic route for aerobic degradation of halobenzoates, halophenol, halobenzenes and haloanilines. 1 = 1,2-dioxygenase, 2 = monooxygenase, 3 = 3,5-cyclohexadiene-1,2-diol-carboxylate dehydrogenase, 4 = catechol-1,2-dioxygenase, 5 = catechol-2,3-dioxygenase, 6 = muconatecycloisomerase, 7 = muconolactoneisomerase, 8 = dienelactonehydrolase.

yielding halocatechols. *Ortho* substituted benzoates give 3-halocatechols, *meta* substitution gives 3- or 4-halocatechols and *para* substitution gives 4-halocatechols. Ring cleavage of these compounds takes place by the *ortho* route, yielding halo-*cis, cis*-muconates (Fig. 4). Many strains capable of degrading halobenzoates have been described. They are members of the genera *Pseudomonas* (Reineke & Knackmuss 1980; Schreiber et al. 1980; Chatterjee et al. 1981; Schmidt & Knackmuss 1984; Focht & Shelton 1987; Wyndham & Straus 1988a; Vora et al. 1988; Hartmann et al. 1989; Schlömann et al. 1990), *Alcaligenes* (Schmidt & Knackmuss 1984; Schmidt 1988; Wyndham & Straus 1988a; Schlömann et al. 1990), *Nocardia* (Cain et al. 1968; Spokes & Walker 1974) and *Azotobacter* (Walker & Harris 1970). Strain FLB 300 which is able to degrade all three monofluorobenzoates was assigned to the *Agrobacterium-Rhizobium* group (Engesser et al. 1990). Bacteria growing on these halobenzoates have catechol-1,2-dioxygenases with high activities towards substituted catechols, referred to as pyrocatechase II. The halogen substituent is eliminated from halo-*cis, cis*-muconate, halomuconolactone or even in the last catabolic step, preceeding the tricarboxylic acid cycle, from halosuccinate (Fig. 4).

Meta cleavage of halocatechols is performed by catechol-2,3-dioxygenases and gives halo-2-hydroxymuconate semialdehydes. These intermediates are toxic to the organisms.

An alternative catabolic route for 3-halobenzoate is via 5-chlorosalicylate and 2,3-dihydroxybenzoate. This route is found in *Bacillus* sp.

(Spokes & Walker 1974) and constructed in *Pseudomonas* WR1 (Lehrbach et al. 1984).

In some cases dehalogenation takes place in the first step. Halogen substitution at the *ortho* position of benzoates can lead to fortuitous oxidative dehalogenation by 1,2-dioxygenases. Oxidative dehalogenation of 2-fluorobenzoate is described for *Pseudomonas* sp. (Goldman et al. 1967; Milne et al. 1968; Vora et al. 1988), for *Acinetobacter calcoaceticus* (Clarke et al. 1975) and for *Pseudomonas* B13 (Schreiber et al. 1980). 2-Chlorobenzoate was oxidatively dehalogenated by *Pseudomonas* B300 (Sylvestre et al. 1989). *Pseudomonas putida* CLB250 oxidatively dehalogenated 2-FBA, 2-BBA and 2-CBA (Engesser & Schulte 1989). In these cases, both 1,2- and 1,6-dioxygenation took place. 1,2-Dioxygenation of 2-halobenzoates leads to catechol and 1,6-dioxygenation to 3-halocatechols. During growth the metabolic route continues via *ortho* cleavage. A mechanism in which chlorine was eliminated in the first step by 2,3-dioxygenase giving 2,3-dihydroxybenzoate as product was found in *Pseudomonas* sp. 2CBA (Fetzner et al. 1989).

Based on growth on salicylate Higson & Focht (1990) suggested hydrolytic dehalogenation yielding salicylate as the first step in 2-halobenzoate degradation by *Pseudomonas aeruginosa* 2-BBZA. However, most reports on hydrolytic dehalogenation involve the displacement of halogen by a hydroxyl group at the *para* position of benzoate. The metabolic route proceeds from 4-halobenzoate to 4-hydroxybenzoate and protocatechuate. 4-Halobenzoate dehalogenation was independent of benzoate degradation. Growing on 4-CBA, *Arthrobacter* sp. DSM 20407 showed *meta* cleavage of protocatechuate but when growing on benzoate the *ortho* cleavage route was followed (Ruisinger et al. 1976). Furthermore, a mutant strain of *Pseudomonas* sp. CBS3, which had lost the ability to grow on 4-CBA, could still grow on benzoate (Keil et al. 1981). The oxygen incorporated was demonstrated to originate from H_2O, not from O_2 (Marks et al. 1984b; Müller et al. 1984).

Genes of *Pseudomonas* sp. CBS3 specifying 4-chlorobenzoate dehalogenase were cloned in *Pseudomonas putida* KT2440. A 9.5 kilobase-pair fragment inserted in a plasmid conferred on this strain the ability to grow on 4-CBA but did not complement mutants unable to grow on 4-HBA (Savard et al. 1990). The 4-CBA dehalogenase enzyme of *Pseudomonas* CBS3 showed higher activities in alcohol than in water (Thiele et al. 1988a) and was also able to dehalogenate 4-chloro-dinitrobenzoates and 4-chloro-dinitrophenols (Thiele et al. 1988b).

Dehalogenation by hydroxylation seems very specific for the *para* position of halobenzoates. However, hydroxylation of 3-CBA was reported in 1972 by Johnston et al. Furthermore, *Acinetobacter* sp. 4CB1 showed both *para* and *meta* dehalogenation of 3,4-dichlorobenzoate, but this compound did not serve as growth substrate (Adriaens et al. 1989). Most bacteria which are capable of 4-CBA dehalogenation also dehalogenate 4-BBA and 4-IBA but not 4-FBA. These reactions are described for *Arthrobacter* sp. SU DSM 20407 (Müller et al. 1988), for *Pseudomonas* sp. CBS3 (Thiele et al. 1987) and for *Alcaligenes denitrificans* NTB-1 (Van den Tweel et al. 1986, 1987). Remarkably *Aureobacterium* sp. RHO25 dehalogenates 4-FBA but not 4-CBA (Oltmanns et al. 1989). Furthermore, Marks et al. (1984a) described a 4-chlorobenzoate dehalogenase with activity for 4-CBA, 4-BBA and 4-FBA.

Under anaerobic conditions halobenzoates are reductively dehalogenated. The first reports on this subject were in 1983 from Horowitz et al. and Suflita et al. They described the reductive dehalogenation of iodo-, bromo- and chlorobenzoates by a methanogenic bacterial consortium isolated from sewage sludge. The dechlorination of chlorobenzoates seemed specific for *meta* substituents whereas dehalogenation of iodo- and bromo-benzoates occurred at all three positions. Fluorobenzoates were never found to be reductively dehalogenated.

Several attempts were made to isolate and characterize the bacterium responsible for this reaction (Shelton & Tiedje 1984; Dolfing & Tiedje 1986; DeWeerd et al. 1986; Dolfing & Tiedje 1987; Stevens et al. 1988; Linkfield & Tiedje 1990; Mohn et al. 1990;) De Weerd et al. (1990) named this bacterium *Desulfomonile tiedjei* strain DCB-1 now catalogued as (ATCC 49306). This bacterium is a sulphate reducer and can use also 3-chlorobenzoate as terminal electron acceptor in the absence of sulphate. The reduction of 3-chlorobenzoate is coupled to ATP production (Dolfing 1990), which results in an increase of growth yield (Mohn & Tiedje 1990) and is thus a new form of anaerobic respiration.

A denitrifying consortium was also found to dehalogenate chlorobenzoates (Sharak Genthner et al. 1989). The dehalogenation occurred in presence of nitrate.

Reductive dechlorination was described for *Alcaligenes denitrificans* NTB-1Y under aerobic circumstances (van den Tweel et al. 1987). Prior to *para* hydroxylation, 2,4-dichlorobenzoate was reductively dehalogenated at the *ortho* position. Groenewegen et al. (1990) later characterized this bacterium as *Coryneform bacterium* NTB-1 and showed that ATP was necessary for the transport of 4-chlorobenzoate through the cell membrane.

Degradation of halogenated benzenes

Marinucci & Bartha (1979) described the degradation of 1,2,3- and 1,2,4-trichlorobenzenes to CO_2 in soil and cultures inoculated with soil. They identified 3,4,5-tri-, 2,6-di- and 2,3-dichlorophenol as metabolites of 1,2,3-TrCBz and 2,4-, 2,5- and 3,4-dichlorophenol in incubations with 1,2,4-TrCBz. Similarly, Ballschmiter & Scholz (1981) isolated 2,3-, 3,4- and 2,6-dichlorophenols from incubations of three *Pseudomonas putida* strains with 1,2-DCBz and 2,4,6-DCP from 1,3,5-TrCBz.

In contrast, the chlorobenzene-utilizing strain WR1306 was shown to degrade this compound via dioxygenation to form 3-chlorocatechol, which is further degraded by the *ortho*-cleavage pathway (Fig. 4) (Reineke & Knackmuss 1984). Reineke and Knackmuss suggested that the chlorophenols isolated by other workers were artefacts produced by acid-catalyzed dehydration of the *cis*-dihydrodiols formed by dioxygenation. In common with other chloroaroma-degrading bacteria, this strain appeared to contain ring cleavage enzymes showing high activity towards chlorinated substrates.

A similar pathway is utilized by two *Alcaligenes* strains which are able to grow on 1,3- and 1,4-DCBz, chlorobenzene and benzene (de Bont et al. 1986; Schraa et al. 1986), by *Pseudomonas* strain JS6 able to degrade all three dichlorobenzenes (Spain & Nishino 1987), by a 1,2-DCBz-utilising *Pseudomonas* strain (Haigler et al. 1988) and by *Pseudomonas* strain P51 which can grow on 1,2,4-TrCBz and the three dichlorobenzenes (van der Meer et al. 1987). Pyrocatechases (catechol-1,2-oxidases) with high activities towards chlorinated substrates also appear to be induced in such strains grown on chlorobenzenes. Although methyl-substituted benzenes are degraded via the *meta*-cleavage pathway, a mutant of strain JS6 degrades 4-chlorotoluene via the *ortho*-pathway (Haigler & Spain 1989).

Kröckel & Focht (1987) constructed a chlorobenzene-utilizing recombinant *Pseudomonas putida* strain from mixed cultures of toluene-grown *Pseudomonas putida* and benzoate-grown *Pseudomonas alcaligenes* strains exposed to chlorobenzene. Chromosomal DNA from the *Pseudomonas alcaligenes* strain was transferred and integrated in a TOL-like plasmid of the *Pseudomonas putida*. During insertion a 24 kB fragment was lost from the plasmid, which resulted in the loss of the ability to grow on xylene and methylbenzoates. This fragment coded for a meta-cleavage pyrocatechase with low specifity and high activity (Carney et al. 1989a, b).

Recently, bromobenzene-utilizing *Pseudomonas* strains have been isolated from chemostat cultures exposed to increasing concentrations of bromobenzene (Sperl & Harvey 1988). Bromocatechols appeared to be intermediates in the degradation of this compound.

Reductive dechlorination is the dominant degradative reaction of chlorobenzenes under anaerobic conditions. Hexachlorobenzene was dechlorinated by two pathways in anaerobic sewage sludge (Fathepure et al. 1988). The major route gave pentachlorobenzene, 1,2,3,5-TCBz and 1,3,5-TrCBz whereas the minor route yielded PCBz, 1,2,4,5-

TCBz, 1,2,4-TrCBz and dichlorobenzenes. There was no evidence for further reduction of 1,3,5-TrCBz. In contrast, Bosma et al. (1988) observed reductive dechlorination of all tri- and dichlorobenzene isomers in anaerobic sediment columns.

Degradation of halogenated phenols

Bacteria able to grow on pentachlorophenol, the most commonly used chlorophenol, were first described in the early 1970s (Chu & Kirsch 1972; Watanabe 1973). Three pathways have been identified for the biodegradation of chlorophenols and other halophenols. Mono- and di-chlorophenols are oxygenated to chlorocatechols, whereas the higher chlorinated phenols are hydroxylated to form chlorinated hydroquinones. Under anaerobic conditions chlorophenols undergo initial reductive dechlorination.

The 3-chlorobenzoate-utilizing *Pseudomonas* B 13 is also able to grow on phenol and 4-chlorophenol and to degrade other mono- and dichlorophenols (Knackmuss & Hellwig 1978). The chlorophenol-degrading bacteria show high activity of pyrocatechase II, a catechol 1,2-dioxygenase able to cleave chlorocatechols. The pathway proposed for the degradation of chlorophenols consists of initial monooxygenation to form chlorocatechols, which undergo *ortho* ring cleavage to chloromuconic acids, lactonization with loss of chloride and further degradation (the β-ketoadipate pathway). Cometabolism of 2-, 3- and 4-chloro- and 2,4- and 3,4-dichlorophenols by *Nocardia* sp. DSM 43251 follows a similar mechanism (Engelhardt et al. 1979).

Degradation of monochlorophenols was poor in defined mixed bacterial cultures containing *Pseudomonas* and *Alcaligenes* strains (Schmidt et al. 1983), due to the accumulation of toxic metabolites formed by *meta* cleavage of chlorocatechols. In the presence of *Pseudomonas* B 13, which degrades chlorocatechols via *ortho* cleavage, the chlorophenols were degraded with release of chloride. Hybrid strains were isolated from such mixed cultures which were able to grow on all three monochlorophenols, for example *Alcaligenes* strain A 7-2 (Schwien & Schmidt 1982).

The 2,4,5-trichlorophenoxyacetic acid-degrading *Pseudomonas* cepacia strain AC1100 is able to degrade a range of di-, tri-, tetra- and pentachlorophenols (Karns et al. 1983a). These chlorophenols are wholly or partly dechlorinated. Dehalogenation was also observed of 2,4-di-, 2,4,6-tri- and penta-bromophenol, but not of 2,4,6-triiodophenol. The enzymes responsible for dechlorination of 2,4,5-T, 2,4,5-trichlorophenol and PCP are induced by 2,4,5-TrCP (Karns et al. 1983b). The mechanism by which 2,4,5-TrCP is degraded by this strain has been identified as conversion of 2,4,5-TrCp to 2,5-dichlorohydroquinone, which undergoes a dehalogenation to 5-chloro-2-hydroxyhydroquinone and subsequent ring cleavage (Sangodkar et al. 1989).

Rhodococcus strains An 117 and An 213 co-metabolize monochlorophenols via the β-ketoadipate pathway (Janke et al. 1988a). Ring cleavage is catalyzed by 'ordinary' catechol-1,2-dioxygenases with low activity towards chlorocatechols. Degradation of 3- and 4-CP, but not of 2-CP, is stimulated in the presence of glucose as an extra source of energy and reducing equivalents (Janke et al. 1988b).

In contrast, the PCP-utilizing *Rhodococcus chlorophenolicus* initially attacks tri-, tetra- and pentachlorophenols by *para*-hydroxylation to produce chlorinated hydroquinones (Apajalahti & Salkinoja-Salonen 1987a). The hydroxyl group is derived from a water molecule. However, this reaction only takes place in the presence of molecular oxygen, which implies the involvement of ATP in the degradation. The tetrachlorohydroquinone formed from PCP is subsequently converted to a dichlorotrihydroxybenzene by a reaction involving both hydrolytic and reductive dechlorinations (Apajalahti & Salkinoja-Salonen 1987b) (Fig. 5). Two further reductive dechlorinations then give 1,2,4-trihydroxybenzene. Trichlorohydroquinone is degraded very slowly, suggesting that it is not an intermediate in this pathway. Similar results have been found for *Rhodococcus* strain CP-2 (Häggblom et al. 1988; 1989a, b).

A pathway involving initial hydrolytic dechlorination in the *para* position to form tetrachlorohydroquinone and further reductive dechlorinations was responsible for the degradation of PCP by an aerobic *Flavobacterium* strain (Steiert & Crawford 1986). This strain is also able to degrade and dechlorinate a range of di-, tri- and tetra-chlorophenols (Steiert et al. 1987). Chlorophenols with chlorine substituents in both *ortho* (2 and 6) positions were degraded most readily. Of these, 2,4,6-TrCP, 2,3,5,6-TeCP and PCP were inducers of the complete PCP degradation pathway.

Many microorganisms, including several strains of *Rhodococcus, Acinetobacter* and *Pseudomonas*, are able to *O*-methylate halophenols (Allard et al. 1985; Neilson et al. 1988; Häggblom et al. 1989).

Reductive dechlorination of mono-, di- and pentachlorophenols and pentabromophenol takes place in anaerobic sewage sludges (Boyd & Shelton 1984; Mikesell & Boyd 1986). Which positions are dechlorinated most rapidly depend on which monochlorophenol the sludges are adapted to. Complete reductive dehalogenation of PCP and PBP and mineralization to methane and CO_2 was observed in sludges adapted to all three monochlorophenols. Reductive dehalogenation of 2- and 3-CP, 2,4-DCP, 2,4-DBP and 2,4,6-TrBP has also been observed in anaerobic consortia enriched from aquatic sediments (King 1988; Sharak Genthner et al. 1989a,b).

Degradation of halogenated anilines

Chlorinated anilines are formed by the degradation of many pesticides in the environment. In the presence of nitrate-reducing bacteria, they undergo condensations to chlorinated azobenzenes, triazenes and biphenyls (Minard et al. 1977; Corke et al. 1979). However, these appear to be chemical reactions of diazonium cations derived from the chloroanilines. The role of the bacteria is reduction of nitrate to nitrite, which reacts with chloroanilines to form diazonium cations. Such reactions take place for a variety of substituted anilines, including mono-and dichloroanilines, but not trichloroanilines, and for monobromoanilines, but not for 2-fluoroaniline (Lammerding et al. 1982).

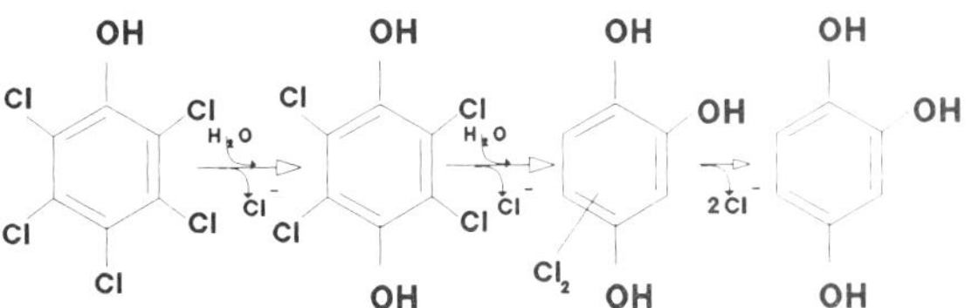

Fig. 5. Metabolic route for PCP degradation by *Rhodococcus chlorophenolicus*.

Few bacteria are known that can mineralize halogenated anilines. *Moraxella* sp. strain G is able to use aniline, 4-fluoro-, 2-chloro-, 3-chloro-, 4-chloro- and 4-bromoanilines, but not 4-iodoaniline, as sole carbon and nitrogen source (Zeyer & Kearney 1982a; Zeyer et al. 1985). This strain is also able to co-metabolize 2,4-DCA (Zeyer & Kearney 1982b; Zeyer et al. 1985). Degradation of these compounds proceeds by initial dioxygenation to form halocatechols, catalyzed by an aniline oxidase with a broad substrate specificity. Further degradation was by a modified *ortho*-cleavage pathway involving a catechol-1,2-oxidase with high activity towards substituted catechols (Zeyer et al. 1985).

Similar pathways are involved in the cometabolism of monochloroanilines by two *Rhodococcus* sp. strains (Janke et al. 1988a, b) and 3,4-DCA by a *Pseudomonas putida* strain (You & Bartha 1982). A *Pseudomonas* strain which degrades aniline via the *meta*-cleavage pathway is not able to degrade chlorinated anilines, although these compounds do induce the enzymes for aniline oxidation (Konopka et al. 1989).

Reductive dehalogenation of chloroanilines takes place under anaerobic conditions. In methanogenic, but not sulphate-reducing, aquifers sequential *ortho* and *para* dehalogenation of 2,3,4,5-TCA yielded 2,3,5-TrCA and 3,5-DCA (Kuhn & Suflita 1989). 3,4-DCA was dechlorinated to 3-CA. No further dehalogenations were detected.

Degradation of halogenated phenoxyacetic acids

The biodegradation of the herbicides 2,4-dichlorophenoxyacetic acid (2,4-D) and 2,4,5-trichlorophenoxyacetic acid (2,4,5-T) has been investigated by various groups. In general, biodegradation of 2,4-D takes place via initial cleavage of the ether bond, followed by hydroxylation of the resulting dichlorophenol to chlorocatechols (e.g. Bollag et al. 1968a,b; Evans et al. 1971; Tiedje & Alexander 1969).

One of the best characterized 2,4-D-degrading microorganisms is *Alcaligenes eutrophus* JMP 134. This strain also degrades 4-chloro-2-methylphenoxyacetic acid, 2-methylphenoxyacetic acid and phenoxyacetic acid (Pieper et al. 1988). Ester bond cleavage of these compounds is apparently catalyzed by a monooxygenase with a wide substrate specificity. For the chlorinated compounds, the chlorocatechol intermediates are cleaved by the *ortho* mechanism but non-chlorinated compounds are cleaved by both *ortho* and *meta* routes. *Flavobacterium* strain MH degrades not only 2,4-D by this pathway, but also a range of other 2,4-dichlorophenoxyalkanoic acids (Horvath et al. 1990).

As mentioned above, the 2,4,5-T-utilizing *Pseudomonas cepacia* AC1100 initially converts this compound to 2,4,5-TrCP (Karns et al. 1983a), which is then dechlorinated to form 2,5-dichlorohydroquinone (Sangodkar et al. 1989). This strain degrades 2,4-D to chlorohydroquinone, which accumulates and inhibits 2,4,5-T degradation (Haugland et al. 1990). The 2,4-D-degrading *Alcaligenes eutrophus* strain JMP134 does not degrade 2,4,5-T. In mixed cell suspensions of strains AC1100 and JMP134 exposed to both 2,4-D and 2,4,5-T, chlorohydroquinone and chlorophenols accumulate, similarly to that in pure suspensions of AC1100. Presumably, the 2,4-D-degradation pathway of JMP134 cannot compete with that of AC1100. Conjugative transfer of the plasmid coding for 2,4-D degradation in JMP1344 to AC1100 gave a constructed strain able to simultaneously degrade 2,4-D and 2,4,5-T.

Reductive dehalogenation of chlorophenoxyacetic acids appears to be an important reaction under anaerobic conditions. 2,4,5-T was dehalogenated in methanogenc aquifer samples to form 2,4- and 2,5-dichlorophenoxyacetic acids (Gibson & Suflita 1990). Further degradation resulted in the formation of monochlorophenoxyacetic acids, chlorophenols and phenol. These reactions were inhibited by added sulphate, but stimulated by added organic substrates.

Degradation of halogenated biphenyls

The degradation of chlorinated biphenyls has been reviewed by Furukawa (1982), Parsons et al. (1983) and Safe (1984). The ability to degrade polychlorinated biphenyls (PCBs) aerobically is found in several genera of both Gram positive and Gram negative bacteria (Ohmori et al. 1973; Walia et al. 1988; Untermann et al. 1988). These are mostly members of the genera *Pseudomonas, Alcaligenes, Arthrobacter* and *Acinetobacter*.

Many reports describe the mineralization (i.e complete degradation to CO_2, often measured by the formation of $^{14}CO_2$) of individual chlorinated biphenyls (Shiaris & Sayler 1982; Kong & Sayler 1983; Bailey et al. 1983; Fries & Marrow 1984; Brunner et al. 1985) brominated biphenyls (Kong & Sayler 1983) and commercial PCB mixtures (Hankin & Sawhney 1984; Baxter & Sutherland 1984; Brunner et al. 1985). However, in many cases chlorinated benzoates accumulated.

Mineralization rates are enhanced by sunlight (Kong & Sayler 1983) and moderately aerobic circumstances (Pardue et al. 1988). Substrate enrichment and inoculation with PCB-degrading bacteria enhanced also mineralization (Brunner et al. 1985).

Mineralisation capabilities differ in different bacterial consortia due to differing metabolic abilities of the members of the population (Hiramoto et al. 1989; Pettigrew et al. 1990). One possible explanation is that biphenyls are initially degraded by *meta* cleavage whereas *meta* cleavage of the resulting chlorobenzoates leads to toxic end products (see above). Only when enzymes for *ortho* cleavage are also induced will complete mineralization be possible. *Alcaligenes* strain JB1 is able to co-metabolize both chlorobiphenyls and chlorobenzoates (Parsons et al. 1988). Complete mineralization of 4-chlorobiphenyl by a two-membered culture of *Pseudomonas* CBS3 and a facultative

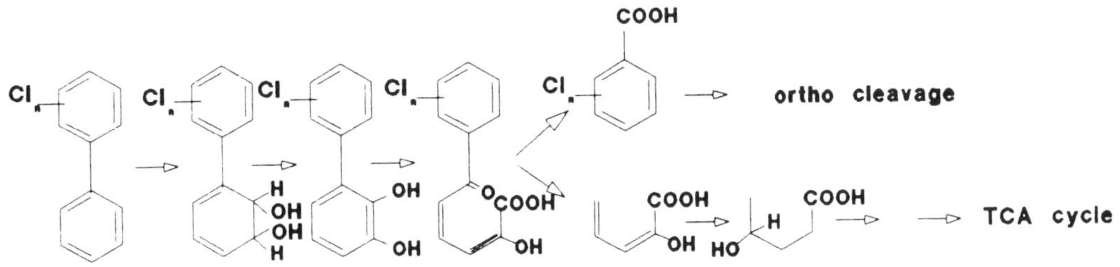

COOH
Cl ortho cleavage
H OH OH H OH OH COOH COOH COOH H HO OH TCA cycle

Fig. 6. Major metabolic route for aerobic degradation of PCBs.

anaerobic strain B-206 was described by Sylvestre et al. (1985). Also a coculture of *Acinetobacter* sp. strain P6 and *Acinetobacter* strain 4CB1 mineralizes 3,4-, 4,4′ DCBP and 3,3′,4,4′-TCBP by initial hydrolytic dehalogenation of the chlorobenzoate intermediates (Adriaens et al. 1989; Adriaens & Focht 1990).

The major aerobic microbial degradation pathway of PCBs, consists of 2,3-dioxygenation of the less substituted aromatic ring, *meta* cleavage and further degradation to chlorobenzoates (Fig. 6). The elimination of chlorine is thought to be a fortuitous event, which occurs in later metabolic steps. Based on studies of PCB metabolism by *Alcaligenes* sp. Y46 and *Acinetobacter* sp. P6 Furukawa (1982) proposed the following relationships between PCB structure and biodegradability.

- The less chlorinated the biphenyl, the faster aerobic degradation takes place. Biphenyls with more than 5 chlorines substituted are resistant to degradation.
- Dioxygenation takes place on the ring with the least chlorine substituents.
- Nonchlorinated vincinal *ortho* and *meta* positions favour dioxygenation.
- PCBs with chlorine substituents on both rings are more recalcitrant than isomers containing an unchlorinated ring.
- Congeners with substituted *ortho* positions are recalcitrant.

In accordance with these results *Alcaligenes* strain JB1 showed fast degradation of 2,2′,3,3′-tetrachlorobiphenyl and relatively slow degradation of 3,3′,4,4′-TCB (Parsons et al. 1988). Also cultures of *Pseudomonas* sp. KKS102 to which supernatant from a culture of a related *Pseudomonas* sp. KKS101 had been added showed relatively fast degradation of 2,2′,3,3′-TCB and slow degradation of 2,2′6,6′-TCB and 3,3′,4,4′-TCB (Kimbara et al. 1988). Similar results were described for *Corynebacterium* sp. MB1 (Bedard et al. 1986).

However, in contrast to these results *Alcaligenes eutrophus* H850 and *Pseudomonas putida* LB400 completely metabolize 2,2′,5,5′-TCB and even degrade 2,2′,4,4′,6,6′-HCB (Bedard et al. 1986). Furthermore, *Alcaligenes eutrophus* H850 does not degrade 2,2′,3,3′-TCB whereas *Pseudomonas putida* LB400 does degrade this latter compound but does not metabolize 2,2′,6,6′-TCB (Bedard et al. 1986,1987a; Bopp 1986). Dioxygenation at the 3,4-position was proposed for the degradation of 2,2′,5,5′-TCB by these organisms. All the *meta* cleavage pathway enzymes were detected in strain H850 (Unterman et al. 1988). A new metabolite, 2,4,5-trichloroacetophenone was detected in incubations of *Alcaligenes eutrophus* H850 with 2,2′,4,4′,5,5′-HCB (Bedard et al. 1987b). Some other PCB congeners may also be degraded via chloroacetophenone intermediates.

Nitration of 4-chlorobiphenyl has been reported for strain B-206 (Sylvestre et al. 1982).

The first reports on anaerobic PCB degradation involved the analysis of PCBs in sediment samples. Compared to the chlorinated aromatic mixtures originally discharged, decreased relative concentrations of highly chlorinated biphenyls were detected in river sediments. Comparison of changes in PCB concentrations in sterile and non-sterile sediments demonstrated that the bacterial community was responsible for reductive dehalogenation of PCBs (Brown et al. 1987a,b; Quensen et al. 1988,1990). Differences in congener specificity were detected in different consortia. The final products of anaerobic dechlorination are the *ortho* substituted congeners, which can be degraded aerobically. Thus, total degradation of PCBs seems possible by sequential anaerobic and aerobic treatment. However, recently it was shown that dehalogenation only occurs with biphenyls with up to seven chlorines. More highly chlorinated congeners were not dehalogenated (Quensen et al. 1990).

Degradation of halogenated dibenzo-*p*-dioxins and dibenzofurans

Very little is known of the degradation of these compounds. The co-metabolism of mono-, di- and trichlorodioxins by a biphenyl-utilizing *Beijerinckia* strain has been described (Klecka & Gibson 1980). *cis*-1,2-Dihydrodiols were isolated as the products of dioxygenation of 1-chloro- and 2-chlorodioxins. Acid-catalyzed dehydration of these compounds gives 2-hydroxylated compounds. Further metabolism of the dihydrodiols produces 1,2-dihydroxylated derivatives, but there was no evidence for ring cleavage of these compounds. In fact, the 1,2-dihydroxylated derivatives appear to inhibit the ring cleavage enzymes in this strain. The biphenyl-utilizing *Alcaligenes* strain JB1 appears to co-metabolize mono-, di- and trichlorinated dioxins by the same mechanism (Parsons & Storms 1989).

Very slow oxidative degradation of 2,3,7,8-tetrachlorodibenzo-*p*-dioxin (2,3,7,8-TCDD) has been reported for a number of microorganisms, including *Pseudomonas testosteroni, Bacillus megaterium* and *Nocardiopsis* strains (Philippi et al. 1982; Quensen et al. 1983). Traces of polar metabolites, probably hydroxylated derivatives, are formed by these strains.

To date no evidence for ring cleavage of chlorinated dibenzo-*p*-dioxins has been reported. Recently, however, oxidative ring cleavage of dibenzo-*p*-dioxin by a dibenzofuran-degrading *Pseudomonas* strain was described (Harms et al. 1990).

There are almost no reports of the biodegradation of halogenated dibenzofurans. Strubel et al. (1989) stated that dibenzofuran-degrading cultures metabolized chlorinated dibenzofurans, but gave no details. Recently, degradation of 2-CDF and 2,8-DCDF by *Alcaligenes* strain JB1 was reported (Parsons et al. 1990).

References

Adriaens P, Kohler H-PE, Kohler-Staub D & Focht DD (1989) Bacterial dehalogenation of chlorobenzoates and coculture biodegradation of 4,4′-dichlorobiphenyl. Appl. Environ. Microbiol. 55: 887–892

Adriaens P & Focht DD (1990) Continuous coculture degradation of selected polychlorinated biphenyl congeners by *Acinetobacter* spp. in an aerobic reactor system. Environ. Sci. Technol. 24: 1042–1049

Allard A-S, Remberger M & Neilson AH (1985) Bacterial O-methylation of chloroguaiacols: Effect of substrate concentration, cell density, and growth conditions. Appl. Environ. Microbiol. 49: 279–288

Apajalatiti JHA & Salkinoja-Salonen MS (1987a) Dechlorination and para-hydroxylation of polychlorinated phenols by *Rhodococcus chlorophenolicus*. J. Bacteriol. 169: 675–681

Apajalatiti JHA & Salkinoja-Salonen MS (1987b) Complete dechlorination of tetrachlorohydroquinone by cell extracts of pentachlorophenol-induced *Rhodococcus chlorophenolicus*. J. Bacteriol. 169: 5125–5130

Bailey RE, Gonsior SJ & Rhinehart WL (1983) Biodegradation of the monochlorobiphenyls and biphenyl in river water.Environ. Sci. Technol. 17: 617–624

Ballschmiter K & Scholz C (1981) Primärschritte der Umwandlung von Chlorbenzol-Derivaten durch *Pseudomonas putida*. Angew. Chem. 93: 1026–1027

Baxter RM & Sutherland DA (1984) Biochemical and photochemical processes in the degradation of chlorinated biphenyls. Environ. Sci. Technol. 18: 608–610

Bedard DL, Unterman R, Bopp LH, Brennan MJ, Haberl ML & Johnson C (1986) Rapid assay for screening and characterizing microorganisms for the ability to degrade polychlorinated biphenyls. Appl. Environ. Microbiol. 51: 761–768

Bedard DL, Wagner RE, Brennan MJ, Haberl ME & Brown Jr JF (1987a) Extensive degradation of Arochlors and environmentally transformed polychlorinated biphenyls by *Alcaligenes eutrophus* H850. Appl. Environ. Microbiol. 53: 1094–1102

Bedard DL, Haberl ML, May RJ & Brennan MJ (1987b) Evidence for novel mechanisms of polychlorinated biphenyl metabolism in *Alcaligenes eutrophus* H850. Appl. Environ. Microbiol. 53: 1103–1112

Bollag JM, Helling CS & Alexander M (1968a) 2,4-D Metabolism: Enzymatic hydroxylation of chlorinated phenols. J. Agric. Food Chem. 16: 826–828

Bollag JM, Briggs GG, Dawson JE & Alexander M (1968b) 2,4-D Metabolism: Enzymatic degradation of chlorocatechols. J. Agric. Food Chem. 16: 829–833

de Bont JAM, Vorage MJAW, Hartmans S & van den Tweel

WJJ (1986) Microbial Degradation of 1,3-Dichlorobenzene. Appl. Environ. Microbiol. 52: 677–680

Bopp LH (1986) Degradation of highly chlorinated PCBs by *Pseudomonas* strain LB400. J. Ind. Microbiol. 1: 23–29

Bosma TNP, Van der Meer JR, Schraa G, Tros ME & Zehnder AJB (1988) Reductive dechlorination of all trichloro- and dichlorobenzene isomers. FEMS Microbiol. Ecol. 53: 223–229

Boyd SA & Shelton DR (1984) Anaerobic biodegradation of chlorophenols in fresh and acclimated sludge. Appl. Environ. Microbiol. 47: 272–277

Brown Jr JF, Bedard DL, Brennan MJ, Carnahan JC, Feng H & Wagner RE (1987a) Polychlorinated biphenyl dechlorination in aquatic sediments. Science 236: 709–712

Brown Jr JF, Wagner RE, Feng H, Bedard DL, Brennan MJ, Carnahan JC & May RJ (1987b) Environmental dechlorination of PCBs. Environ. Toxicol. Chem. 6: 579–593

Brunner W, Sutherland FH & Focht DD (1985) Enhanced biodegradation of polychlorinated biphenyls in soil by analog enrichment and bacterial inoculation. J. Environ. Qual. 14: 324–328

Cain RB, Trantner EK & Darrah JA (1968) The utilization of some halogenated aromatic acids by Nocardia: oxidation and metabolism. Biochem. J. 106: 211–227

Carney BF, Kröckel L, Leary JV & Focht DD (1989a) Identification of *Pseudomonas alcalignes* chromosomal DNA in the plasmid DNA of the chlorobenzene-degrading recombinant *Pseudomonas putida* strain CB1-9. Appl. Environ. Microbiol. 55: 1037–1039

Carney BF & Leary JV (1989b) Novel alterations in plasmid DNA associated with aromatic hydrocarbon utilization by *Pseudomonas putida* R5-3. Appl. Environ. Microbiol. 55: 1523–1530

Chatterjee DK, Hamada S & Chakrabarty AM (1981) Plasmid specifying total degradation of 3-chlorobenzoate by a modified *ortho* pathway. J. Bacteriol. 146: 639–646

Chu JP & Kirsch EJ (1972) Metabolism of pentachlorophenol by an axenic bacterial culture. Appl. Environ. Microbiol. 23: 1033–1035

Clarke KF, Callely AG, Livingstone A & Fewson CA (1975) Metabolism of monofluorobenzoates by *Acinetobacter calcoaceticus* N.C.I.B.8250: formation of monofluorocatechols. Biochim. Biophys. Acta 404: 169–179

Corke CT, Bunce NJ, Beaumont AL & Merrick RL (1979) Diazonium cations as intermediates in the microbial transformation of chloroanilines to chlorinated biphenyls, azo compounds, and triazenes. J. Agric. Food Chem. 27: 644–646

DeWeerd KA, Suflita JM, Linkfield T, Tiedje JM & Pritchard PH (1986) The relationship between reductive dehalogenation and other aryl substituent removal reactions catalyzed by anaerobes. FEMS Microbiol. Ecol. 38: 331–339

DeWeerd KA, Mandelco L, Tanner RS, Woese CR & Suflita JM (1990) *Desulfomonile tiedjei* gen nov and sp nov, a novel anaerobic dehalogenating sulfate-reducing bacterium. Arch. Microbiol. 154: 23–30

Dolfing J & Tiedje JM (1986) Hydrogen cycling in a three-tiered food web growing on the methanogenic conversion of 3-chlorobenzoate. FEMS Microbiol. Ecol. 38: 293–298

(1987) Growth yield increase linked to reductive dechlorination in a defined 3-chlorobenzoate degrading methanogenic coculture. Arch. Microbiol. 149: 102–105

Dolfing J (1990) Reductive dechlorination of 3-chlorobenzoate is coupled to ATP production and growth in an anaerobic bacterium, strain DCB-1. Arch. Microbiol. 153: 264–266

Engelhardt G, Rast HG & Wallnöfer PR (1979) Cometabolism of phenol and substituted phenols by *Nocardia* spec. DSM 43251. FEMS Microbiol. Lett. 5: 377–383

Engesser KH & Schulte P (1989) Degradation of 2-bromo-, 2-chloro- and 2-fluorobenzoate by *Pseudomonas putida* CLB 250. FEMS Microbiol. Lett. 60: 143–148

Engesser KH, Auling G, Busse J & Knackmuss H-J (1990) 3-Fluorobenzoate enriched bacterial strain FLB 300 degrades benzoate and all three isomeric monofluorobenzoates. Arch. Microbiol. 153: 193–199

Evans WC, Smith BSW, Fernley HN & Davies JI (1971) Bacterial Metabolism of 2,4-Dichlorophenoxyacetate. Biochem. J. 122: 543–552

Fathepure BZ, Tiedje JM & Boyd SA (1988) Reductive dechlorination of hexachlorobenzene to tri- and dichlorobenzenes in anaerobic sewage sludge. Appl. Environ. Microbiol. 54: 327–330

Fetzner S, Müller R & Lingens F (1989) A novel metabolite in the microbial degradation of 2-chlorobenzoate. Biochem. Biophys. Res. Commun. 161: 700–705

Focht DD & Shelton D (1987) Growth kinetics of *Pseudomonas alcaligenes* C-0 relative to inoculation and 3-chlorobenzoate metabolism in soil. Appl. Environ. Microbiol. 53: 1846–1849

Fries GF & Marrow GS (1984) Metabolism of chlorobiphenyls in soil. Bull. Environ. Contam. Toxicol. 33: 6–12

Furukawa K (1982) Microbial degradation of polychlorinated biphenyls (PCBs). In: Chakrabarty AM (Ed) Biodegradation and Detoxification of Environmental Pollutants CRC Boca Raton FLA (pp 33-57)

Gibson SA & Suflita JM (1990) Anaerobic degradation of 2,4,5-trichlorophenoxyacetic acid in samples from methanogenic aquifer: Stimulation by short-chain organic acids and alcohols. Appl. Environ. Microbiol. 56: 1825–1832

Goldman P, Milne GWA & Pignataro MT (1967) Fluorine containing metabolites formed from 2-fluorobenzioc acid by *Pseudomonas* species. Arch. Biochem. Biophys. 118: 178–184

Groenewegen PEJ, Driessen AJM, Konings WN & de Bont JAM (1990) Energy-dependent uptake in the *Coryneform bacterium* NTB-1. J. Bacteriol. 172: 419–423

Haigler BE, Nishino SF & Spain JC (1988) Degradation of 1,2-dichlorobenzene by a *Pseudomonas* sp. Appl. Environ. Microbiol. 54: 294–301

Haigler BE & Spain JC (1989) Degradation of *p*-chlorotoluene by a mutant of *Pseudomonas* sp. strain JS6. Appl. Environ. Microbiol. 55: 372–379

Häggblom MM, Nohynek LJ & Salkinoja-Salonen MS (1988) Degradation and O-methylation of chlorinated phenolic compounds by *Rhodococcus* and *Mycobacterium* strains. Appl. Environ. Microbiol. 54: 3043–3052

Häggblom MM, Janke D & Salkinoja-Salonen MS (1989a) Hydroxylation and dechlorination of tetrachlorohydroquinone by *Rhodococcus* sp. strain CP-2 cell extracts. Appl. Environ. Microbiol. 55: 516–519

Häggblom MM, Janke D, Middeldorp PJM & Salkinoja-Salonen MS (1989b) O-Methylation of chlorinated phenols in the genus *Rhodococcus*. Arch. Microbiol. 152: 6–9

Hankin L & Sawhney BL (1984) Microbial Degradation of Polychlorinated Biphenyls in Soil. Soil Sci. 137: 401–407

Harms H, Wittich R-M, Sinnwell V, Meyer H, Fortnagel P & Francke W (1990) Transformation of dibenzo-*p*-dioxin by *Pseudomonas* sp. strain HH69. Appl. Environ. Microbiol. 56: 1157–1159

Hartmann J, Reineke W & Knackmuss H-J (1979) Metabolism of 3-chloro-, 4-chloro-, and 3,5-dichlorobenzoate by a Pseudomonad. Appl. Environ. Microbiol. 37: 421–428

Hartmann J, Engelberts K, Nordhaus B, Schmidt E & Reineke W (1989) Degradation of 2-chlorobenzoate by in vivo constructed hybrid Pseudomonads. FEMS Microbiol. Lett. 61: 17–22

Haugland RA, Schlemm DJ, Lyons III RP, Sferra PR & Chakrabarty AM (1990) Degradation of the chlorinated phenoxyacetate herbicides 2,4-dichlorophenoxyacetic acid and 2,4,5-trichlorophenoxyacetic acid by pure and mixed bacterial cultures. Appl. Environ. Microbiol. 56: 1357–1362

Higson FK & Focht DD (1990) Degradation of 2-Bromobenzoic Acid by a Strain of *Pseudomonas aeruginosa*. Appl. Environ. Microbiol. 56: 1615–1619

Hiramoto M, Ohtake H & Toda K (1989) A kinetic study on total degradation of 4-chlorobiphenyl by a two-step culture of *Arthrobacter* and *Pseudomonas* strains. J. Fermentation Bioeng. 1: 68–70

Horowitz A, Suflita JM & Tiedje JM (1983) Reductive dehalogenations of halobenzoates by anaerobic lake sediment microorganisms. Appl. Environ. Microbiol. 45: 1459–1461

Horvath M, Ditzelmüller G, Loidl M & Streichsbier F (1990) Isolation and characterization of a 2-(2,4-dichlorophenoxy) propionic acid-degrading soil bacterium. Appl. Microbiol. Biotechnol. 33: 213–216

Janke D, Al-Mofarji T, Straube G, Schumann P & Prauser H (1988a) Critical steps in the degradation of chloroaromatics by *Rhodococci*. I. Initial enzyme reactions involved in catabolism of aniline, phenol and benzoate by *Rhodococcus* sp. An 117 and An 213. J. Basic Microbiol. 8: 509–518

Janke D, Al-Mofarji T & Schukat B (1988b) Critical steps in degradation of chloroaromatics by *Rhodococci*. II. Whole-cell turnover of different monochloroaromatic non-growth substrates by *Rhodococcus* sp. An 117 and An 213 in the absence/presence of glucose. J. Basic Microbiol. 8: 519–528

Johnston HW, Briggs GG & Alexander M (1972) Metabolism of 3-chlorobenzoic acid by a *Pseudomonad*. Soil. Biol. Biochem. 4: 187–190

Karns JS, Kilbane JJ, Duttagupta S & Chakrabarty AM (1983a) Metabolism of halophenols by 2,4,5-trichlorophenoxyacetic acid-degrading *Pseudomonas cepacia*. Appl. Environ. Microbiol. 46: 1176–1181

Karns JS, Duttagupta S & Chakrabarty AM (1983b) Regulation of 2,4,5-trichlorophenoxyacetic acid and chlorophenol metabolism in *Pseudomonas cepacia* AC 1100. Appl. Environ. Microbiol. 46: 1182–1186

Keil H, Klages U & Lingens F (1981) Degradation of 4-chlorobenzoate by *Pseudomonas* sp. CBS3: induction of catabolic enzymes. FEMS Microbiol. Lett. 10: 213–215

Kimbara K, Hashimoto T, Fukuda M, Koana T, Takagi M, Oishi M & Yano K (1988) Isolation and characterization of a mixed culture that degrades polychlorinated biphenyls. Agric. Biol. Chem. 52: 2885–2891

King GM (1988) Dehalogenation in marine sediments containing natural sources of halophenols. Appl. Environ. Microbiol. 54: 3079–3085

Klecka GM & Gibson DT (1980) Metabolism of dibenzo-*p*-dioxin and chlorinated dibenzo-*p*-dioxins by a *Beijerinckia* species. Appl. Environ. Microbiol. 39: 288–296

Knackmuss H-J & Hellwig M (1978) Utilization and cooxidation of chlorinated phenols by *Pseudomonas* sp. B13. Arch. Microbiol. 117: 1–7

Kong H-Y & Sayler GS (1983) Degradation and total mineralization of monohalogenated biphenyls in natural sediment and mixed microbial culture. Appl. Environ. Microbiol. 46: 666–672

Konopka A, Knight D & Turco RF (1989) Characterization of a *Pseudomonas* sp. capable of aniline degradation in the presence of secondary carbon sources. Appl. Environ. Microbiol. 55: 385–389

Kröckel L & Focht DD (1987) Construction of chlorobenzene-utilizing recombinants by progressive manifestation of a rare event. Appl. Environ. Microbiol. 53: 2470–2475

Kuhn EP & Suflita JM (1989) Sequential reductive dehalogenation of chloroanilines by microorganisms from a methanogenic aquifer. Environ. Sci. Technol. 23: 848–852

Lammerding AM, Bunce NJ, Merrick RL & Corke CT (1982) Structural effects on the microbial diazotization of anilines. J. Agric. Food Chem. 30: 644–647

Lehrbach RP, Zeyer J, Reineke W, Knackmuss H-J & Timmis KN (1984) Enzyme Recruitment in vitro: Use of Clones Genes to Extend the Range of Haloaromatics Degraded by *Pseudomonas* sp. strain B13. J. Bacteriol. 158: 1025–1032

Linkfield TG & Tiedje JM (1990) Characterization of the requirements and substrates for reductive dehalogenation by strain DCB-1. J. Ind. Microbiol. 5: 9–16

Marinucci AC & Bartha R (1979) Biodegradation of 1,2,3- and 1,2,4-trichlorobenzene in soil and in liquid enrichment culture. Appl. Environ. Microbiol. 38: 811–817

Marks TS, Smith ARW & Quirk AV (1984a) Degradation of 4-chlorobenzoic acid by *Arthrobacter* sp. Appl. Environ. Microbiol. 48: 1020–1025

Marks TS, Wait R, Smith ARW & Quirk AV (1984b) The origin of the oxygen incorporated during the dehalogenation/hydroxylation of 4-chlorobenzoic acid by an *Arthrobacter* sp. Biochem. Biophys. Res. Commun. 124: 669–674 van der Meer JR, Roelofsen W, Schraa G & Zehnder AJB (1987) Degradation of Low Concentrations of Dichlorobenzenes and 1,2,4-Trichlorobenzene by *Pseudomonas* sp. P51 in Nonsterile Soil Columns. FEMS Microbiol. Ecol. 45: 333–341

Mikesell MD & Boyd SA (1986) Complete reductive dechlorination and mineralization of pentachlorophenol by anaerobic microorganisms. Appl. Environ. Microbiol. 52: 861–865

Milne GWA, Goldman P & Holtzman JL (1968) The metabolism of 2-fluorobenzoic acid: studies with $^{18}O2$. J. Biol. Chem. 243: 5374–5376

Minard RD, Russel S & Bollag JM (1977) Chemical transformation of 4-chloroaniline to a triazene in a bacterial culture medium. J. Agric. Food Chem. 25: 841–

Mohn WW, Linkfield TG, Pankratz HS & Tiedje JM (1990) Involvement of a collar structure in polar growth and cell division of strain DCB-1. Appl. Environ. Microbiol. 56: 1206–1211

Mohn WM & Tieje JM (1990) Strain DCB-1 conserves energy for growth from reductive dechlorination coupled to formate oxidation. Arch. Microbiol. 153: 267–271

Müller R, Thiele J, Klages U & Lingens F (1984) Incorporation of [$^{18}O\ H_2O$] water into 4-hydroxybenzoic acid in the reaction of 4-chlorobenzoate dehalogenase from *Pseudomonas* spec. CBS3. Biochem. Biophys. Res. Commun. 124: 178–182

Müller R, Oltmans RH & Lingens F (1988) Enzymic dehalogenation of 4-chlorobenzoate by extracts from *Arthrobacter* sp. SU DSM 20407. Biol. Chem. Hoppe-Seyler 369: 567–571

Neilson AH, Lindgren C, Hynning P-A & Remberger M (1988) Methylation of halogenated phenols and thiophenols by cell extracts of Gram-positive and Gram-negative bacteria. Appl. Environ. Microbiol. 54: 524–530

Ohmori T, Ikai T, Minoda Y & Yamada K (1973) Utilization of Hydrocarbons by Microorganisms. XXV. Utilization of Polyphenyl and Polyphenyl-related Compounds by Microorganimsms. Agric. Biol. Chem. 37: 1599–1605

Oltmanns RH, Müller R, Otto MK & Lingens F (1989) Evidence for a new pathway in the bacterial degradation of 4-fluorobenzoate. Appl. Environ. Microbiol. 55: 2499–2504

Pardue JH, Delaune RD & Patrick Jr. WH (1988) Effect of sediment pH and oxidation-reduction potential on PCB mineralization. Water Air Soil Pollut. 37: 439–447

Parsons J, Veerkamp W & Hutzinger O (1983) Microbial metabolism of chlorobiphenyls. Toxicol. Environ. Chem. 6: 327–350

Parsons JR, Sijm DTHM, van Laar A & Hutzinger O (1988) Biodegradation of chlorinated biphenyls and benzoic acids by a *Pseudomonas* strain. Appl. Microbiol. Biotechnol. 29: 81–84

Parsons JR & Storms MCM (1989) Biodegradation of chlorinated dibenzo-p-dioxins in batch and continuous cultures of strain JB1. Chemosphere 19: 1297–1308

Parsons JR, Ratsak C & Siekerman C (1990) Biodegradation of chlorinated dibenzofurans by an *Alcaligenes* strain. In: Hutzinger O and Fiedler H (Eds) Organohalogen Compounds. Proc. Dioxin '90 – EPRI Seminar, Sept. 10–14, 1990, Bayreuth, Vol 1 (pp 377–380). Ecoinforma Press, Bayreuth, F.R.G.

Pettigrew CA, Breen A, Corcoran C & Sayler GS (1990) Chlorinated Biphenyl Mineralization by Individual Populations and Consortia of freshwater Bacteria. Appl. Environ. Microbiol. 56: 2036–2045

Philippi M, Schmid J, Wipf HK & Hütter RA (1982) A microbial metabolite of TCDD. Experientia 38: 659–661

Pieper DH, Reineke W, Engesser K-H, Knackmuss H-J (1988) Metabolism of 2,4-dichlorophenoxyacetic acid, 4-chloro-2-methylphenoxyacetic acid and 2-methylphenoxyacetic acid by *Alcaligenes eutrophus* JMP 134

Quensen III JF & Matsumura F (1983) Oxidative degradation of 2,3,7,8-tetrachlorodibenzo-p-dioxin by microorganisms. Environ. Toxicol. Chem. 2: 261–268

Quensen III JF, Tiedje JM & Boyd SA (1988) Reductive dechlorination of polychlorinated biphenyls by anaerobic microorganisms from sediments. Science 242: 752–754

Quensen III JF, Boyd SA & Tiedje JM (1990) Dechlorination of Four Commercial Polychlorinated Biphenyl Mixtures (Aroclors) by Anaerobic Microorganisms from Sediments. Appl. Environ. Microbiol. 56: 2360–2369

Reineke W (1984) Microbial degradation of halogenated aromatic compounds. Microbiol. Ser. 13: 319–360

Reineke W & Knackmuss H-J (1984) Microbial metabolism of haloaromatics. Isolation and properties of a chlorobenzene-degrading bacterium. Appl. Environ. Microbiol. 47: 395–402

Reineke W & Knackmuss H-J (1988) Microbial degradation of haloaromatics. Ann. Rev. Microbiol. 42: 263–287

Reineke W & Knackmuss H-J (1980) Hybrid pathway for chlorobenzoate metabolism in *Pseudomonas* sp. B13 derivatives. J. Bacteriol. 142: 467–473

Ruisinger S, Klages U & Lingens F (1976) Abbau der 4-Chlorobenzoesaure durch eine *Arthrobacter*species. Arch. Microbiol. 110: 253–256

Safe SH (1984) Microbial degradation of polychlorinated biphenyls. Microbiol. Ser. 13: 361–369

Sangodkar UMX, Aldrich TL, Haugland RA, Johnson J, Rothmel RK, Chapman PJ & Chakrabarty AM (1989) Molecular basis of biodegradation of chloroaromatic compounds. Acta Biotechnol. 9: 301–316

Savard P, Péloquin L & Sylvestre M (1990) Cloning of *Pseudomonas* strain CBS3 Genes Specifying Dehalogenation of 4-Chlorobenzoate. J. Bacteriol. 168: 81–85

Schlömann M, Fischer P, Schmidt E & Knackmuss H-J (1990) Enzymatic Formation, Stability, and Spontaneous Reactions of 4-Fluoromuconolactone, a Metabolite of the Bacterial Degradation of 4-Fluorobenzoate. J. Bacteriol. 172: 5119–5129

Schmidt E (1988) Bioconversion of 3-chlorobenzoate to 2-chloromuconate controlled by on line HPLC. Appl. Microbiol. Biotechnol. 27: 347–350

Schmidt E, Hellwig M & Knackmuss H-J (1983) Degradation of chlorophenols by a defined mixed microbial community. Appl. Environ. Microbiol. 46: 1038–1044

Schmidt E & Knackmuss H-J (1984) Production of *cis, cis-*

muconate from benzoate and 2-fluoro-*cis, cis*-muconate from 3-fluorobenzoate by 3-chlorobenzoate degrading bacteria. Appl. Microbiol. Biotechnol. 20: 351–355

Schraa G, Boone ML, Jetten MSM, van Neerven ARW, Colberg PJ & Zehnder AJB (1986) Degradation of 1,4-dichlorobenzene by *Alcaligenes* sp. strain A175. Appl. Environ. Microbiol. 52: 1374–1381

Schreiber A, Hellwig M, Dorn E, Reineke W & Knackmuss H-J (1980) Critical reactions in fluorobenzoic acid degradation by *Pseudomonas* sp B13. Appl. Environ. Microbiol. 39: 58–67

Schwien U & Schmidt E (1982) Improved degradation of monochlorophenols by a constucted strain. Appl. Environ. Microbiol. 44: 33–39

Sharak Genthner BR, Price II WA & Pritchard PH (1989a) Anaerobic degradation of chloroaromatic compounds in aquatic sediments under a variety of enrichment conditions. Appl. Environ. Microbiol. 55: 1466–1471

(1989b) Characterization of anaerobic dechlorinating consortia derived from aquatic sediments. Appl. Environ. Microbiol. 55: 1472–1476

Shelton DR & Tiedje JM (1984) Isolation and partial characterization of bacteria in an anaerobic consortium that mineralizes 3-chlorobenzoic acid. Appl. Environ. Microbiol. 48: 840–848

Shiaris MP & Sayler GS (1982) Biotransformation of PCBs by natural assemblages of freshwater microorganisms. Environ. Sci. Technol. 16: 367–369

Spain JC & Nishino SF (1987) Degradation of 1,4-dichlorobenzene by a *Pseudomonas* sp. Appl. Environ. Microbiol. 53: 1010–1019

Sperl GT & Harvey GJ (1988) Microbial adaptation to bromobenzene in a chemostat. Curr. Microbiol. 17: 99–103

Spokes JR & Walker N (1974) Chlorophenol and chlorobenzoic acid co-metabolism by different genera of soil bacteria. Arch. Microbiol. 96: 125–134

Steiert JG & Crawford RL (1986) Catabolism of pentachlorophenol by a *Flavobacterium* bacterium. Biochem. Biophys. Res. Commun. 141: 825–830

Steiert JG, Pignatello JJ & Crawford RL (1987) Degradation of chlorinated phenols by a pentachlorophenol-degrading bacterium. Appl. Environ. Microbiol. 53: 907–910

Stevens TO, Linkfield TG & Tiedje JM (1988) Physiological Characterization of Strain DCB-1, a Unique Sulfidogenic Bacterium. Appl. Environ. Microbiol. 54: 2938–2943

Strubel V, Rast HG, Fietz W, Knackmuss H-J & Engesser KH (1989) Enrichment of dibenzofuran utilizing bacteria with high co-metabolic potential towards dibenzodioxin and other anellated aromatics. FEMS Microbiol. Lett. 58: 233–238

Suflita JM, Robinson JA & Tiedje JM (1983) Kinetics of microbial dehalogenation of haloaromatic substrates in methanogenic environments. Appl. Environ. Microbiol. 45: 1466–1473

Sylvestre M, Mailhiot K, Ahmad D & Massé R (1989) Isolation and preliminary characterization of a 2-chlorobenzoate degrading *Pseudomonas*. Can. J. Microbiol. 35: 439–443

Sylvestre M, Massé R, Ayotte C, Messier F & Fauteux J (1985) Total biodegradation of 4-chlorobiphenyl (4-CB) by a two-membered bacterial culture. Appl. Microbiol. Biotechnol. 21: 192–195

Sylvestre M, Massé R, Messier F, Fauteux J, Bisaillon J-G & Beaudet R (1982) Bacterial nitration of 4-chlorobiphenyl. Appl. Environ. Microbiol. 44: 871–877

Thiele J, Müller R & Lingens F (1987) Initial characterization of 4-chlorobenzoate dehalogenase from *Pseudomonas* sp. CBS3. FEMS Microbiol. Lett. 41: 115–119

(1988a) Enzymatic dehalogenation of 4-chlorobenzoate by 4-chlorobenzoate dehalogenase from *Pseudomonas* sp. CBS3 in organic solvents. Appl. Microbiol. Biotechnol. 27: 577–580

(1988b) Enzymatic dehalogenation of chlorinated nitroaromatic compounds. Appl. Environ. Microbiol. 54: 1199–1202

Tiedje JM & Alexander M (1969) Enzymatic Cleavage of the Ether Bond of 2,4-Dichlorophenoxyacetate. J. Agric. food Chem. 17: 1080–1084

van den Tweel WJJ, Ter Burg N, Kok JB & De Bont JAM (1986) Bioformation of 4-hydroxybenzoate from 4-chlorobenzoate by *Alcaligenes denitrificans* NTB-1. Appl. Microbiol. Biotechnol. 25: 289–294

van den Tweel WJJ, Kok JB & De Bont JAM (1987) Reductive dechlorination of 2,4-dichlorobenzoate to 4-chlorobenzoate and hydrolytic dehalogenation of 4-chloro-, 4-bromo-, and 4-iodobenzoate by *Alcaligenes denitrificans* NTB-1. Appl. Environ. Microbiol. 53: 810–815

Unterman R, Bedard DL, Brennan MJ, Bopp LH, Mondello FJ, Brooks RE, Mobley DP, McDermott JB, Schwartz CC & Dietrich DK (1988) Biological approaches for polychlorinated biphenyl degradation. Basic Life Sciences 45: 253–269

Vora KA, Singh C & Modi VV (1988) Degradation of 2-fluorobenzoate by a *Pseudomonad*. Curr. Microbiol. 17: 249–254

Walia S, Tewari R, Brieger G, Thimm V & McGuire T (1988) Biochemical and genetic characterization of soil bacteria degrading polychlorinated biphenyl. In: Abbou R (Ed) Hazardous Waste: Detection Control Treatment (pp 1621–1632). Elsevier Amsterdam

Walker N & Harris D (1970) Metabolism of 3-chlorobenzoic acid by *Azotobacter* species. Soil Biol. Biochem. 2: 27–32

Watanabe I (1973) Isolation of pentachlorophenol decomposing bacteria from soil. Soil Sci. Plant Nutr. 19: 109–116

Wyndham RC & Straus NA (1988a) Chlorobenzoate catabolism and interaction between *Alcaligenes* and *Pseudomonas* species from Bloody Run Creek. Arch. Microbiol. 150: 230–236

Wyndham RC, Singh RK & Straus NA (1988b) Catabolic instability, plasmid gene deletion and recombination in *Alcaligenes* sp. BR60. Arch. Microbiol. 150: 237–243

You I-S & Bartha R (1982) Cometabolism of 3,4-dichloroaniline by *Pseudomonas putida*. J. Agric. Food Chem. 30: 274–277

Zeyer J & Kearney PC (1982a) Microbial degradation of parachloroaniline as sole carbon and nitrogen source. Pesticide Biochem. Physiol. 17: 215–223

(1982b) Microbial metabolism of propanil and 3,4-dichloroaniline. Pesticide Biochem. Physiol. 224: 231–biodegradation

Zeyer J, Wasserfallen A & Timmis KN (1985) Microbial mineralization of ring-substituted anilines through an *ortho*-cleavage pathway. Appl. Environ. Microbiol. 50: 447–453

Novel Biodegradable Microbial Polymers

Proceedings of the NATO Advanced Research Workshop on New Biosynthetic Biodegradable Polymers of Industrial Interest from Microorganisms, Sitges, Spain, May 26–31, 1990

edited by **Edwin A. Dawes,** *Dept. of Applied Biology, University of Hull, UK*

NATO ADVANCED SCIENCE INSTITUTES SERIES
E: Applied Sciences 186

New Publication

This book provides a detailed account of the latest researches in the field of new biosynthetic, biodegradable polymers of industrial interest that are derived from microorganisms. It stems from a NATO Advanced Research Workshop that brought together polymer chemists, biochemists, microbiologists, geneticists, materials chemists, physicists and engineers who are working on the multidisciplinary aspects of the development of biosynthetic, biocompatible and biodegradable polyesters, polysaccharides, polyphosphates and polysulfides produced by microbes. It is an area of intense international interest; the properties and characteristics of these polymers, the molecular biology and enzymology of their synthesis and, where appropriate, their industrial production and applications, are considered by experts. The relevance of these biodegradable polymers to amelioration of the pollution of the environment currently caused by petrochemical plastics is a theme which underlies many of the contributions. Here, then, is a volume which presents an up-to-date survey and discussion of topics that dominate current research and thinking on microbial polymers.

Contents

Polyesters I: Polymer Characteristics and Properties. **Polyesters II:** Polymer Production – Properties. **Polyesters III:** Molecular Biology and Enzymology. Polyphosphates. Natural Distribution of Microbial Polymers and Sulfur. Polysaccharides I. Polysaccharides II. University, Industry and Government: Future Directions and Priorities in Research and Development – Panel Discussion. Posters: Polyhydroxyalkanoates. Polysaccharides. Index.

1990, 480 pp. ISBN 0–7923–0949–9
Hardbound $142.00/Dfl. 230.00/£79.00

KLUWER ACADEMIC PUBLISHERS

P.O. Box 322, 3300 AH Dordrecht, The Netherlands
P.O. Box 358, Accord Station, Hingham, MA 02018-0358, U.S.A.